Fatouma Mohamed Abdoul-Latif
Ayoub Ainane
Tarik Ainane

ALGAE: FROM MICRO TO MACRO

Fatouma Mohamed Abdoul-Latif
Ayoub Ainane
Tarik Ainane

ALGAE: FROM MICRO TO MACRO

BIODIVERSITY, INNOVATION AND SUSTAINABLE APPLICATIONS IN INDUSTRY

ScienciaScripts

Imprint

Cover image: www.ingimage.com

This book is a translation from the original published under ISBN 978-620-2-54007-0.

Publisher:
Sciencia Scripts
is a trademark of
Dodo Books Indian Ocean Ltd. and OmniScriptum S.R.L publishing group

120 High Road, East Finchley, London, N2 9ED, United Kingdom
Str. Armeneasca 28/1, office 1, Chisinau MD-2012, Republic of Moldova, Europe
Managing Directors: Ieva Konstantinova, Victoria Ursu
info@omniscriptum.com

Printed at: see last page
ISBN: 978-620-8-57150-4

PREFACE :

Algae, the fascinating organisms that populate our seas and rivers, have long been underestimated despite their immense potential. In "**ALGAE: FROM MICRO TO MACRO - BIODIVERSITY, INNOVATIONS, AND SUSTAINABLE APPLICATIONS IN INDUSTRY**", we explore the world of algae in depth, from their incredible diversity to their potential for innovation and sustainable applications. This book was born out of a passion for these unique organisms and the conviction that their thorough understanding and responsible exploitation can play a crucial role in sustainable development. By offering a comprehensive overview, we aim to enlighten readers on the fundamental aspects of algae, their classification, and their uses in various sectors.

The general introduction introduces the subject by presenting algae as key players in biodiversity and industrial innovation. We lay the foundations for the discussion by highlighting their ecological and economic importance.

In Chapter 1: General information on algae, we provide an overview of these organisms, their biology and their role in ecosystems. This chapter is designed to provide a solid foundation on which to build for the more specific discussions that follow.

Chapter 2: Microalgae; focuses on these microscopic forms, often invisible to the naked eye, but whose potential in terms of biotechnology, nutrition and water treatment is exceptional. We explore their characteristics, their applications, and the challenges associated with their cultivation.

Chapter 3: Macro-algae; looks at the larger, more visible algae, such as kelp and red algae. We discuss their biodiversity, their traditional and modern uses, and their impact on the food and pharmaceutical industries.

Chapter 4: Innovations in algae cultivation; explores recent advances in algae cultivation techniques, including new methods that promise to transform their production and use. This chapter highlights key innovations that open up new prospects for the industry.

Finally, Chapter 5: Analyses & Perspectives; offers a reflection on future trends and the challenges to be overcome. We analyze the impact of algae on industry and the environment, and discuss the opportunities and challenges associated with their sustainable use.

This book is aimed at a wide audience, from researchers and professionals to students and environmental enthusiasts. We hope it will contribute to a better understanding of algae and a more responsible use of their potential.

We invite you to dive into the fascinating world of algae, discover their remarkable biodiversity, and explore the innovations shaping their future in industry.

Natural microalgae on the shores of Djibouti.

ALGAE: FROM MICRO TO MACRO - BIODIVERSITY, INNOVATION AND SUSTAINABLE APPLICATIONS IN INDUSTRY

Authors :

Dr. Fatouma Mohamed Abdoul-Latif

Dr. Ayoub Ainane

Dr. Tarik Ainane

CONTENTS :

GENERAL INTRODUCTION

Algae are much less well known than land plants, and are considerably more complex to understand. Mainly found in aquatic, marine and underwater environments, they form an extremely diverse group of organisms, making uniform presentation difficult. Many, indeed the vast majority, are unicellular (microalgae), requiring elaborate microscopic techniques for identification. Systematically, algae are also highly diverse, reflecting their long genetic history. They do not form a homogeneous group among plants, but can be divided into several independent evolutionary lineages. For seaweeds, there are three main evolutionary paths: the brown-yellow line with brown algae, the red line with red algae, and the green line with green algae, mosses, ferns and flowering plants. Despite morphological convergences, there are more genetic differences between a brown alga such as *Fucus* (seaweed) and a green alga of the *Ulva* genus (sea lettuce), than between the latter and an oak tree. This great heterogeneity explains the richness and biochemical diversity of algae, which is why they are so widely exploited.

But how many species of algae are there? The answer is complex, especially for unicellular algae, since their number is so high, their diversity poorly understood, and their census and classification constantly evolving. The international Algae Base database lists some 127,000 species names, most of them microalgae. The number of macro-algae species is estimated at around 9,000, of which 1,500 inhabit the seas of Europe, while the total number of micro-algae varies, depending on estimates, from 100,000 to several million.

The development of this immense resource is still in its infancy. Although certain macro-algae have been exploited since antiquity in most maritime countries, notably in the Far East, their worldwide exploitation remains marginal compared to terrestrial plant production: 15 million tonnes of macro-algae (including 13.5 million tonnes in cultivation) and 7 to 10,000 tonnes of micro-algae (around 100% in cultivation), compared to 4 billion tonnes for agricultural production. France and Europe are lagging behind Asian countries, which have developed production techniques that enable them to meet their enormous needs, particularly in terms of food consumption, and to export raw materials worldwide. In France, and particularly in Brittany, there is considerable know-how in aquaculture production, processing and formulation, enabling us to export more and more products.

Microalgae currently account for just 10,000 tonnes per year of controlled cultivation, i.e. excluding microalgae used for wastewater treatment. Compared with the potential resources available in France, this low level of exploitation is mainly due to the lack of industrial maturity of a sector that is still in its infancy. They do, however, play an important role in the food industry, providing gelling agents such as carrageenans and alginates, which have no terrestrial equivalent. Their use in cosmetics and chemistry is also booming, revealing a revival in their industrial valorization. Microalgae, long used as a food source in Chad and Latin America (Spirulina being in fact a cyanobacteria often assimilated to a microalgae), have only been cultivated and exploited industrially for a few decades. Their development has paralleled that of fish and bivalve shellfish aquaculture, their main use being to feed aquaculture larvae. They are also cultivated to produce beta-carotene and various other molecules of interest for human consumption, cosmetics and pharmaceuticals. Their use for energy purposes was not envisaged until the 1980s, after the first oil crisis, and it is only recently that they have again attracted the attention of research and industry as a potential source of biofuel, notably for aviation. Their high lipid content makes them ideal for this type of use, while their high protein and polyunsaturated fatty acid (omega-3) content makes them good candidates for aquaculture feed. However, the major challenge remains to exploit them on a large scale so that they can play a significant role on the world market in the years to come. This is where the work in progress comes in, and this report aims to provide a better understanding of the challenges and issues to be overcome. In terms of production chains, macro-algae and micro-algae differ in a number of respects, leading to significantly different production and, to a lesser extent, value-added chains. These differences concern cultivation methods, essentially in natural marine environments for macro-algae, and in ponds or photobioreactors for micro-algae; harvesting techniques, easy and close to fishing for macro-algae, but technically difficult and close to industrial biotechnology techniques for micro-algae; and processing, classic and close to the agri-food industry for macro-algae, but requiring advanced biorefinery techniques for micro-algae. The same applies to the markets for each sector, which are intrinsically different.

In response to this disparity, we have chosen to deal with these two major sectors in separate chapters on the state of the art, grouping them together only to discuss the priority areas of research to be implemented on a national scale, aligning them along a common value chain for their exploitation.

CHAPTER 1: GENERAL INFORMATION ON ALGAE

1. The current state of the algae industry worldwide :

1.1 Algae at the heart of the living world: from the wealth of exceptional biodiversity to the existence of original natural substances useful to man :

When discussing microalgae and macroalgae in this book, it's essential to remember that both terms belong to a more general concept: that of "algae". Algae are not a single evolutionary group, but a series of organisms belonging to very diverse phylogenetic groups. Algae constitute a vast artificial assemblage of phylogenetically distant organisms, giving rise to a specific discipline: phycology.

Dealing with algae as a whole is no easy task. For example, physiology, biochemistry and molecular biology reveal that groups of algae, apparently very homogeneous in terms of morphology, do not necessarily have a close evolutionary relationship. Red macro-algae, for example, are more closely related to micro-algae than to green macro-algae.

Despite this complexity, an attempt can be made to define algae. Algae can be considered as organisms generally found in wetlands (marine, brackish or freshwater), mostly photosynthetic and possessing chlorophyll a. Algae therefore occupy specific ecological niches, allowing access to light, although some may experience a non-photosynthetic stage during their life cycle. In the marine environment, which is home to a large proportion of algae, they are mainly found in the photic zone, limited to a depth of 100 meters, or even less, depending on water turbidity. Algae are thus limited to the littoral fringe, for those attached to the substrate (macro-algae), and to the photic layer in open water (micro-algae). Algae play a crucial role in the biosphere, helping to maintain its equilibrium as plant biomass. This is particularly true in the nutrient-rich coastal zone, where phytoplankton blooms form, as well as in the vast macro-algae "fields" along the Breton coastline. In oceanic zones, the richness of plant plankton depends on the upwelling of nutrient-rich deep waters.

Algae also display great morphological diversity, manifested through different levels of organization (unicellular and multicellular) and varying sizes: from 1 to 100 microns for phytoplankton cells to several meters for certain macro-algae. This diversity is also reflected in modes of sexual reproduction and various forms of vegetative propagation.

1.2 Algae and metabolic pathways: primary and secondary metabolisms :

Whatever the diversity of algae, most share one universal characteristic: photosynthesis. A photosynthetic organism acts as a converter of light energy into chemical energy, this reaction being catalyzed by photosynthetic pigments. The chemical energy is stored in the form of polysaccharides, which can then be used as rcscrvc or support compounds. The biochemical study of pigments, reserve and support polysaccharides is common to all algae, and sometimes serves as a simplified classification criterion. It has been observed that substances essential to life are metabolized by all microalgae and macroalgae organisms, a process known as primary metabolism. These primary metabolites reveal the unity of the living world; for example, fucoxanthin is the major pigment common to diatom microalgae and macrophyte brown algae. It's worth noting that the particularities of the marine environment sometimes determine the existence of original and widely exploited primary metabolites, such as polyanionic polysaccharides in macro-algae.

Algae live and develop thanks to a complex set of biochemical reactions, some of which lead to the synthesis of metabolites that are not essential for survival, known as secondary metabolites, which are remarkably rich in terms of chemical molecule families. These secondary metabolites reflect the diversity of the living world. In algae, for example, there is a direct correlation between the percentage of nitrogenous secondary metabolites and the degree of evolution of the algae. Unlike dinoflagellates, diatoms very rarely contain toxins that are active in living organisms.

Domoic acid is an interesting case, being one of the few nitrogenous secondary metabolites found in diatoms such as *Pseudonitzchia*, and also present in the macrophyte red alga *Chondria armata*. This compound is used in pharmacology to study degenerative diseases of the central nervous system.

All these observations on algal metabolism point to the similarity of experimental approaches to the study of molecules extracted from algae. All the families of chemical compounds currently known in the living world are represented in algae, with, in addition, entirely original structures, with no known terrestrial equivalent to date.
This text highlights the points of convergence that link microalgae and macroalgae, particularly in terms of scientific approaches and areas of fundamental research. Recent developments in the chemistry of natural marine substances have gone hand in hand with the development of marine biotechnologies applied to algae. For example, the photobioreactor culture of the micro-plants of certain red macrophyte algae should enable the production of halogenated terpene compounds. The technology developed for micro-algae can be transposed to biotechnologies applied to macro-algae. Several research organizations, such as CNRS, INRA and IFREMER, as well as a number of universities, are participating in this pooling effort through various federative actions, such as the creation of the GDR Biochimar and the response to calls for projects within the framework of investments for the future (IDEALG, IEED GREEN STARS).
In the context of this book, this synergy is particularly relevant for exploring the fields of application and potential markets for active molecules derived from algae, offering an opportunity to develop a common economic and commercial perspective.

1.3 Introduction to algae :

Algae are at the origin of life on Earth, an evolution marked by the appearance of bacteria with chlorophyll pigments that initiated photosynthesis, releasing the oxygen necessary for the development of other life forms. From this common ancestor emerged blue, brown, red and green algae, the latter being the only ones to have given rise to higher plants.
Some algae, having lost their pigments, have given rise to organisms now classified as fungi. The term "algae" conceals a wide variety of organisms that can be divided into two main groups.

- ✓ **Microalgae**: Microalgae, of which there are over a million species, consist of one or more cells and generally measure a few tens of microns. Some float or swim with flagella, while others, without flagella, attach themselves to various substrates. Their biotope is extremely varied and sometimes unsuspected, as illustrated by the red microscopic unicellular algae that gives pink flamingos their color. Phytoplankton, made up of micro-algae, play a crucial role in the primary production of oceanic food chains.

- ✓ **Macro-algae**: Macro-algae, of which there are over 30,000 species, have a vegetative apparatus called a thallus, made up of poorly differentiated cells. The thallus may include a fastening system (cramp), a peduncle (stipe), and a blade (frond) of various shapes. Macroalgae can range in size from a few centimeters to over 40 meters, as in the case of the kelp (Macrocystis).

1.4 Algal diversity and distribution :

It's important to note that not all algae are marine plants; they can also be found in freshwater, in the air, on land and even in ice. What's more, not all marine plants are algae, as in the case of posidonia, which produces more oxygen than a forest of the same surface area, and flowers, which distinguishes it from algae.

1.5 Photosynthesis and pigmentation :

Algae have no roots to absorb nutrients; they filter seawater and capture CO_2, which, in the presence of light, enables the production of carbohydrates $(CH_2O)n$ with the release of O_2. Although they breathe by capturing O_2 dissolved in water and releasing CO_2, they produce more O_2 than they consume.

For photosynthesis, algae use pigments such as chlorophyll a, present in all chlorophyllous plants, and chlorophyll b, specific to green plants. These pigments absorb light and transfer the captured energy to chlorophylls a and b, which act as final receptors.

1.6 Light adaptation of algae :

Blue-green algae possess specific pigments, such as phycocyanin, which make them intolerant of sunlight, explaining their inability to leave the aquatic environment, with the exception of lichens. Red algae, which are also intolerant of sunlight, grow at depth or under brown algae, which limits their expansion outside the aquatic environment. Green algae, on the other hand, have adapted to intense light without the need for additional pigments, and left the aquatic environment around 600 million years ago. Brown algae, characterized by brown-yellow pigments such as fucoxanthin, develop at mid- and supralittoral levels, but despite an attempt to establish themselves on land 200 million years ago, they have remained aquatic.

1.7 How can you tell algae from other plants?

The distinction between algae and other plants has evolved over time, and no universal definition has been adopted by all specialists.

- **Distinction in the 19th century: Thallophytes vs. Cormophytes :**

In the 19th century, an old distinction in the plant kingdom was between thallophytes and cormophytes. Algae were then defined as chlorophyllous thallophytes, i.e. chlorophyllous plants without roots, stems or leaves, as opposed to other plants with a cormus, comprising at least roots for mosses and ferns, and stems and leaves for phanerogams. This classification was based on morphological differentiation, but had its limitations, particularly with algae such as characeae, which were not clearly related to thallophytes.

What's more, this definition was too broad, as it also included plants such as mosses, which have rudimentary roots. To solve this problem, these plants were subsequently classified as non-vascular cryptogams, as opposed to vascular cryptogams.

- **Twentieth-century evolution: Classification based on reproductive organs :**

In the 20th century, the evolution of plant systematics, made possible by the use of the microscope, led to a new classification based on the nature of the reproductive organs. Although less practical, this classification was more rigorous. The plant kingdom is now divided into two groups: plants with gametocysts and those with gametangia.

2. Gametocysts and gametangia :

In the case of gametocysts, the envelope of the mother cell becomes the envelope of the gametocyst, and all the contents of the mother cell are transformed into gametes. In the case of gametangia, on the other hand, part of the contents of the mother cell gives rise to gametes, while the other part forms the envelope of the gametangium, which is then multi-cellular.

3. Algal biology :

Algae, or phycophytes (from the Greek *phukos* = algae; *phuton* = plant), are chlorophyllous thallophytes, i.e. organisms capable of photosynthesis, and therefore autotrophic. Typically aquatic, they are the first link in the food chain, producing almost exclusively in the seas and oceans, which cover 7/10 of the earth's surface. Although mainly marine, algae are also found in freshwater (lakes, ponds, streams) and, more rarely, in the air. Algae photosynthesis depends on light, but this is rapidly absorbed by the water, making light insufficient at depths of just a few meters to compensate for losses due to respiration.

For this reason, algae in aquatic environments are limited to a surface zone generally no deeper than 40 to 60 meters. Beyond that, the marine environment is devoid of producers.

There are two main groups within algal populations. The first is made up of species that float or swim in open water, generally microscopic and often unicellular, forming the vegetal and productive part of plankton and phytoplankton (from the Greek *plankton* = wandering). The second group, known as phytobenthos (from the Greek *benthos* = depth), comprises all aquatic organisms living at the bottom of marine, lake or river bodies of water.

Algae are distinguished by their varied colors, due to the presence of pigments that mask chlorophyll to a greater or lesser extent. This characteristic allows algae to be subdivided into three major lineages that are biochemically, structurally and functionally distinct: green algae, brown algae and red algae. Each lineage comprises unicellular species, often considered primitive, and multicellular species of increasing complexity, marked by an increasingly pronounced division of physiological labor. Examination of these plants reveals great plasticity and testifies to numerous evolutionary trials, resulting in a great diversity of organizational plans and modes of reproduction, at the origin of varied development cycles.

The study of algae is of particular interest because of the clues they provide about the stages marking the progression of life forms. Recent molecular biology data indicate that algae are at the origin of all present-day plant groups, probably through an endosymbiosis of a cyanobacterium (chlorophyll prokaryote) that evolved as a plastid in an archaic eukaryotic cell. This common ancestor gave rise to red and green algae, the latter of which gave rise to the embryophytes.

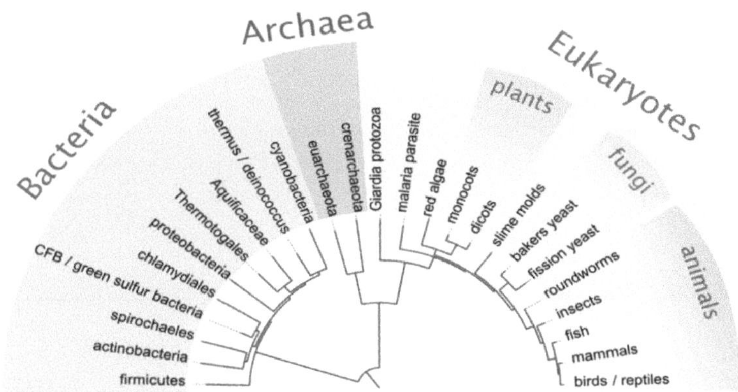

Figure 1: Simplified phylogenetic classification of the living world based on the genome

4. Physiology of algae :

The loss of color underwater is mainly due to the selective absorption of light. Algae, which carry out photosynthesis, absorb light from their environment, whether saltwater or freshwater. When light penetrates water, it undergoes significant changes. Long-wave radiation, such as red, is absorbed first, while shorter-wave radiation, such as blue and green, penetrates more deeply.

Seawater acts as a filter, progressively absorbing light radiation in decreasing order of wavelength, from 700 nm to 400 nm in the visible spectrum. Red radiation disappears completely at depths of around 10 metres. As you go deeper, the other radiations are also absorbed, leaving only green and blue radiations, and only in small quantities, at around 75 to 100 meters depth. This phenomenon explains why certain colors fade rapidly under water, and why algae are limited to shallow areas where light, although filtered, remains sufficiently intense to enable photosynthesis.

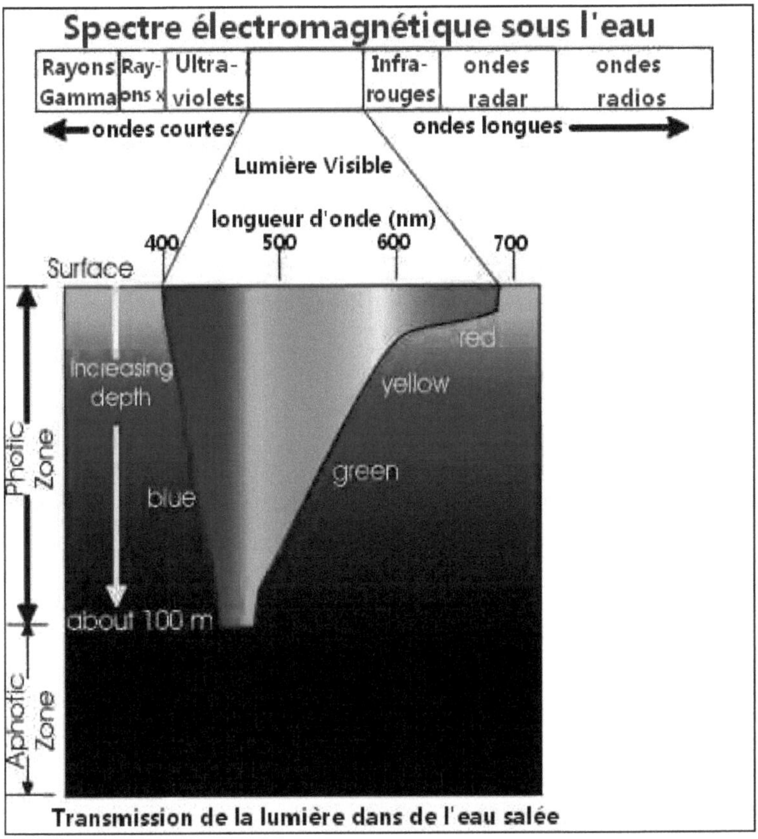

Figure 2: Absorption of different wavelengths of the light spectrum under salt water

5. Selective underwater light absorption :

Visible light, or white light, is made up of a spectrum of colors ranging from violet to red, arranged in ascending order of wavelength. Underwater, light intensity decreases rapidly and exponentially, depending on the wavelength of each component. This phenomenon, known as selective absorption, is caused by the vibrations and deformations of water molecules excited by light absorption.

Higher wavelengths, such as red, are absorbed more strongly, with exact values depending on the transparency of the water, influenced by elements such as sediment and plankton. Red, being the most affected, loses 1/3 of its intensity after one meter, and disappears almost completely after 4 to 5 meters underwater. The accompanying figure illustrates the relative importance of different colors in Mediterranean water, showing the "effective" extinction distance of a color as a function of its wavelength. In the case of sunlight, the distance covered is generally related to depth, which explains the coastal gradation of algae according to color.

6. Light absorption by algae :

Algae depend on light for their development and reproduction, photosynthesizing using pigments in their chloroplasts. These pigments enable them to absorb light, which they use to produce dioxygen and transform light energy into matter, such as sucrose, which forms the walls of algae.
Different algae are distributed at depth according to the "color" of the light (or wavelength) they absorb, thanks to their specific pigmentary equipment. An algae absorbs the complementary color of its own color: a red algae absorbs green radiation, while a green algae absorbs red radiation. Each type of algae has different pigments that enable it to absorb specific wavelengths:

- ✓ **Green algae**: They contain chlorophyll a and b, as well as carotenoids. They absorb mainly blue (400 to 500 nm) and red (around 650 nm) wavelengths, with maximum absorption close to 100 arbitrary units (AU). They absorb little in the green.

- ✓ **Brown algae**: They contain chlorophyll a and c, as well as carotenoids such as fucoxanthin, enabling them to absorb mainly blues and greens.

- ✓ **Red algae**: They contain chlorophyll a and d, carotenoids and specific pigments such as phycoerythrin and phycocyanin. They absorb strongly at blue and green wavelengths (450 to 600 nm), with maximum absorption close to 100 AU. They also absorb in the reds, but less efficiently (around 50 AU for wavelengths above 600 nm).

Cyanobacteria, also known as cyanophyceae or blue-green algae (former names), are photosynthetic bacteria that, like plants, take advantage of the sun's energy to synthesize their organic molecules. To capture this light, they use various pigments, including phycocyanins (blue-green in color) and chlorophyll.

These organisms have developed a set of strategies that have enabled them to colonize various ecosystems, giving them an advantage over algae under certain favorable conditions and allowing them to persist in the environment. Cyanobacteria are distinguished by their ability to adapt to a wide range of environmental conditions, and thrive particularly well in extreme environments. So it's not surprising that they were among the first microorganisms to colonize the Earth, and that they continue to be highly competitive, even dominant, in certain environments.

One of the main assets of cyanobacteria is their diverse pigmentation, which ensures high photosynthetic efficiency. They are capable of sustaining high growth rates, even at low light intensities. It's phycobiliproteins, such as phycocyanin in freshwater, that give cyanobacteria this advantage over phytoplanktonic algae, which rely mainly on chlorophyll a. These pigments enable them to exploit a wider range of solar radiation wavelengths.

In addition, cyanobacteria have developed various protective strategies against ultraviolet (UV) and excessive radiation, enabling them to survive and thrive in conditions of intense light or in environments exposed to high levels of UV radiation. These capacities for adaptation and resistance partly explain their evolutionary success and their dominant presence in many ecosystems today.

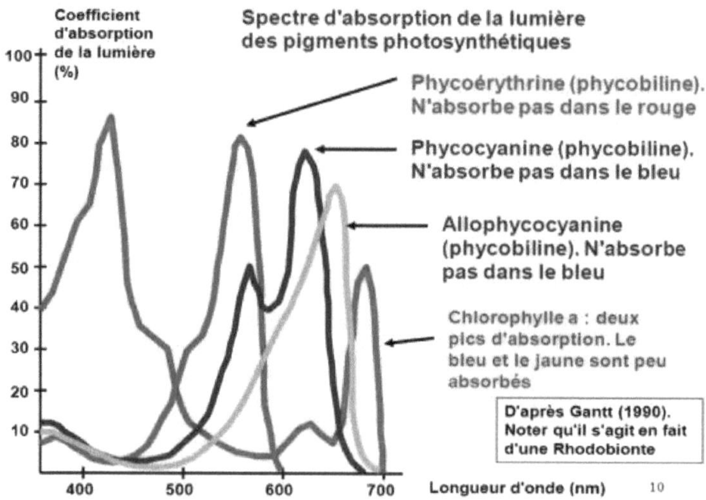

Figure 3: Light absorption spectrum of photosynthetic pigments

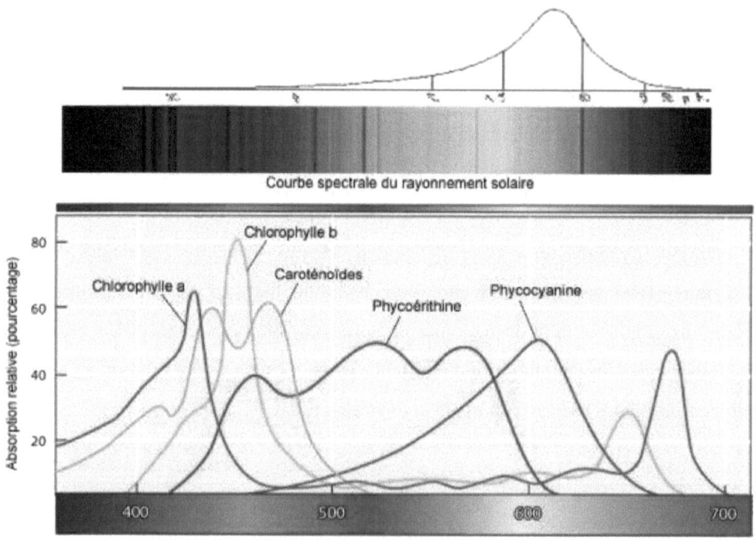

Figure 4: Absorption spectrum of photosynthetic pigments and spectral curve of solar radiation

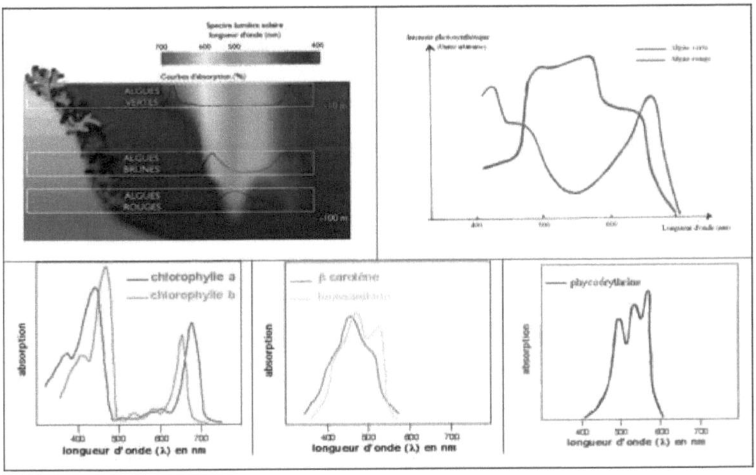

Figure 5: Absorption and action spectra of photosynthetic pigments as a function of water depth

7. Algal ecology :

Ecologically speaking, algae are the first link in the food chain, being the almost exclusive primary producers of the seas and oceans, which cover 7/10 of the earth's surface. They are also present in freshwater, notably in lakes, ponds and streams. Phytoplankton, which form the bulk of the biomass in these environments, play a crucial role, being responsible for most of the planet's oxygen production and carbon dioxide absorption, thanks to photosynthesis.

8. Seaweed spreading on the coast :

The distribution of algae on the coast is influenced by the limit of humectation by sea spray or waves, but depends mainly on light (see section on physiology). This distribution is also affected by the hydrodynamics of the area, i.e. whether or not the area is sheltered from waves. This phenomenon of vertical distribution of algae is known as tiering.

- ✓ **Supralittoral zone**: Located at the interface between sea and land, this zone is never submerged, but receives sea spray. Conditions here are very difficult for life, making it a zone often reduced to a narrow strip colonized mainly by encrusting species, such as certain lichens.
- ✓ **Mediolittoral stage**: Corresponding to the tidal and wave-balancing zone, this stage is limited to a few tens of centimeters in the Mediterranean. The organisms that live here must be able to withstand alternating immersion and emersion.
- ✓ **The infralittoral stage**: Always submerged, this stage represents the zone of maximum biodiversity in the marine environment. The diversity of substrates, whether soft or hard, influences the distribution of species, which vary according to the type of substrate present.

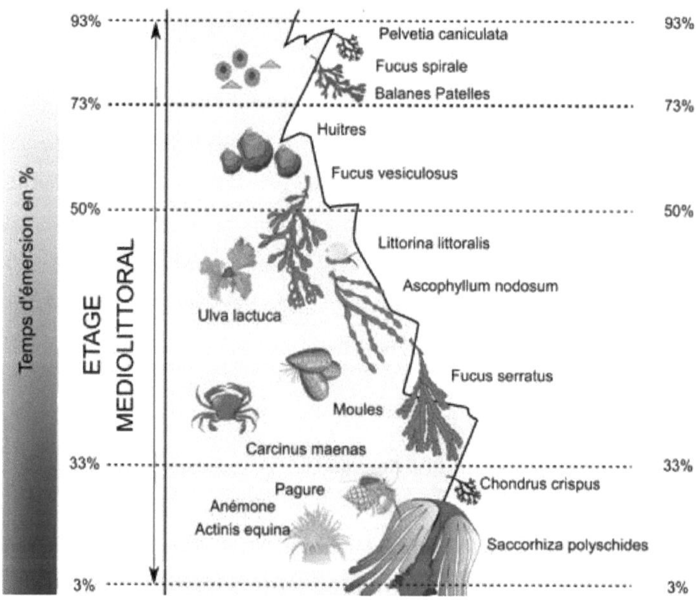

Figure 6: Seaweed distribution along the Atlantic coastline

Figure 7: Seaweed distribution in the Mediterranean infralittoral zone

9. Macroalgae distribution

The geographical and bathymetric distribution of macroalgae is influenced by several factors.

- ✓ **Light**: Light plays a crucial role in the development of macroalgae, varying according to water depth and turbidity. Algae are not found beyond a depth of 50 to 70 meters, becoming rare beyond 30 meters. Their distribution is therefore limited to less than 1% of the ocean surface. Photophilic algae, which require a lot of light, thrive near the surface, while sciaphilic algae, which need less light, thrive at depth, under overhangs or among other algae. The presence of specific pigments in brown and red algae enables them to survive at greater depths than green algae, absorbing blue or violet waves that can no longer be used by chlorophyll for photosynthesis.
- ✓ **Temperature**: Temperature, both aerial and aquatic, is also a determining factor in the distribution of macroalgae, although the thermal inertia of the water attenuates daily and seasonal variations, especially for deep-seated algae.
- ✓ **Nature of the substrate**: The nature of the substrate is another essential factor. Although algae draw their nutrients from seawater rather than soil, substrate texture plays a crucial role, as algae only thrive on hard substrates.
- ✓ **Water Agitation**: Water agitation influences the morphology of thalli within the same species, depending on whether they are found in sheltered or wave-exposed environments.

- ✓ **Drying out** : Drying out, linked to the duration of emersion and the level of establishment on the foreshore, determines the capacity of the algae to retain water.
- ✓ **Salinity**: Salinity fluctuates sharply at low tide, due to evaporation, wind, temperature and precipitation, particularly affecting algae located at the top of the foreshore or in rocky basins, which are said to be euryhaline because of their resistance to these variations.

- ✓ **Algae distribution**: All these factors lead to the distribution of algae in horizontal bands forming distinct belts. However, ocean waters, due to the absence of upwelling water rich in mineral salts, become veritable deserts.
- ✓ **Temperature and phytoplankton**: As on land, marine biological processes proceed more rapidly at higher temperatures. The phytoplankton content of the water therefore depends on latitude and season. However, in tropical and subtropical regions, the difference in temperature between bottom and surface water creates a barrier that traps mineral salts at the bottom, making them inaccessible to microalgae; this explains the lack of phytoplankton in tropical waters. Under ideal conditions, phytoplankton can double their biomass, but these conditions remain limited.

10.The Reproduction of Algae :

The way algae reproduce, and the elements involved, have evolved considerably over the last two billion years, making this subject a crucial aspect of the study of organismal evolution.

- ➢ **Sexual reproduction :**

Sexual reproduction in algae involves specialized cells called gametes. These sex cells are haploid, meaning they possess only one copy of their genetic material. When two gametes from individuals of different sexes meet and fuse, they form an egg or zygote. This then becomes a diploid cell, containing two copies of genetic material: one from the female gamete, the other from the male gamete. By successive division, this cell gives rise to a new organism with half the genetic information of each parent.

Initially, the gametes were morphologically identical (isogamy), but over time the "female" sex cells differentiated (anisogamy). The latter accumulated reserves, becoming larger than the male gametes. At the same time, the reproductive organs, or gametocysts, became specialized. Originally, this organ was simply a wall of the mother cell, which, through division, formed the gametes.

This made it possible to distinguish the oocyst (female reproductive organ) from the spermatocyst (male reproductive organ). The female gamete then remains in the oocyst and attracts male gametes through an expansion called trichogyne, a process known as trichogamy.

11. The reproductive cycles of algae :

Algae have three distinct reproductive cycles:

- ✓ **Monogenetic cycles**: Each thallus is unisexual and haploid, producing a single type of gamete. The meeting of gametes of different sexes produces a diploid zygote that evolves into a thallus, either male or female.
- ✓ **Digenetic cycles**: The diploid egg evolves into a thallus called a meiosporophyte, which produces female and male spores. These spores develop into a unisexual haploid thallus (gametophyte), which produces gametes. Ulva is an example of this cycle.
- ✓ **Trigenetic cycles**: The diploid zygote evolves into a carposporophyte, usually attached like a parasite to the female gametophyte, which gives carpospores that evolve into meiosporophytes. In all these cycles, the haploid gametophyte stage predominates.

Characeae, and *Chara foetida* in particular, represent a hinge group with Bryophytes (mosses), where we can see the beginnings of gametangia and the oosphere or archegonia.

12. Algae life cycle

The life cycle of algae generally takes place in two phases, each characterized by a distinct form. These two forms have more or less marked morphological differences. The sporophyte form produces mature spores, which are released into the environment and disperse. After attaching themselves to the substrate, the spores germinate to give rise to a new organism called a gametophyte, which produces reproductive cells.

Male and female organs may be found on the same thallus or on different individuals. The union of a male and female gamete forms an egg which, after germination, gives rise to a new sporophyte. Thus, the algal life cycle is characterized by the alternation of a sexual phase (gametophyte) and an asexual phase (sporophyte). However, depending on the algae species, this cycle can present numerous variations.

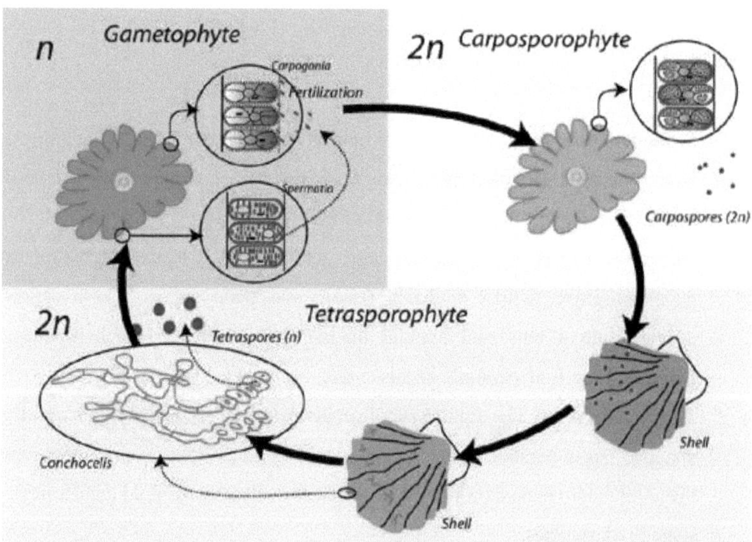

Figure 8: *Porphyra umbilicalis* (Rhodophyceae) cycle: trigenetic or 3 generations

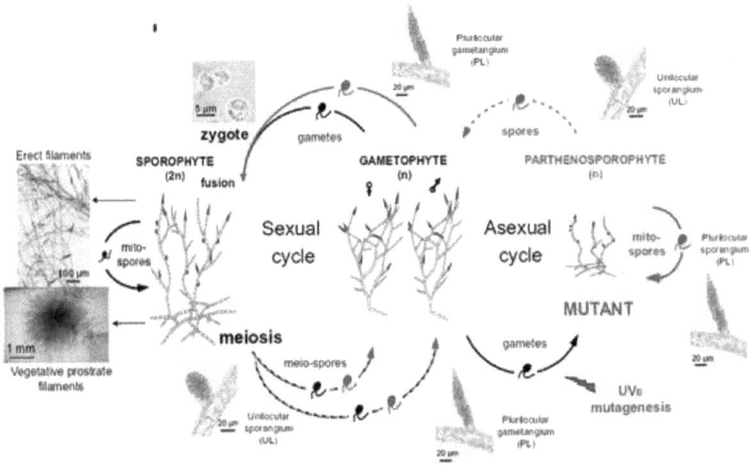

Figure 9: Sexual and asexual reproduction in *Ectocarpus siliculosus* (Phaeophyceae)

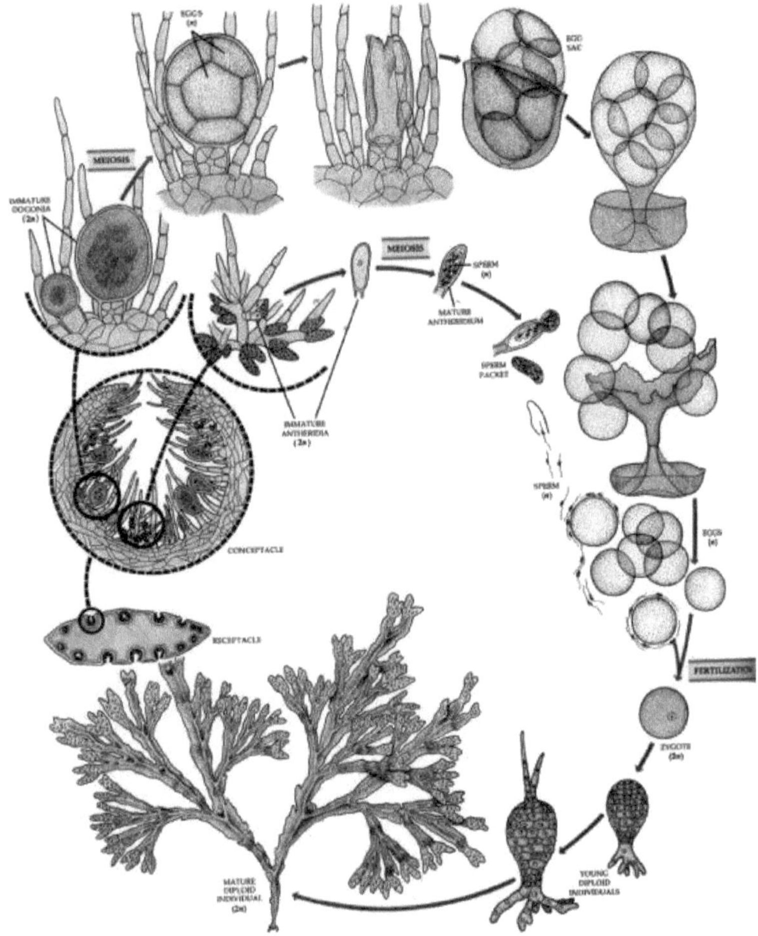

Figure 10: Sexual reproduction in *Fucus vesiculosus* (Phaeophyceae)

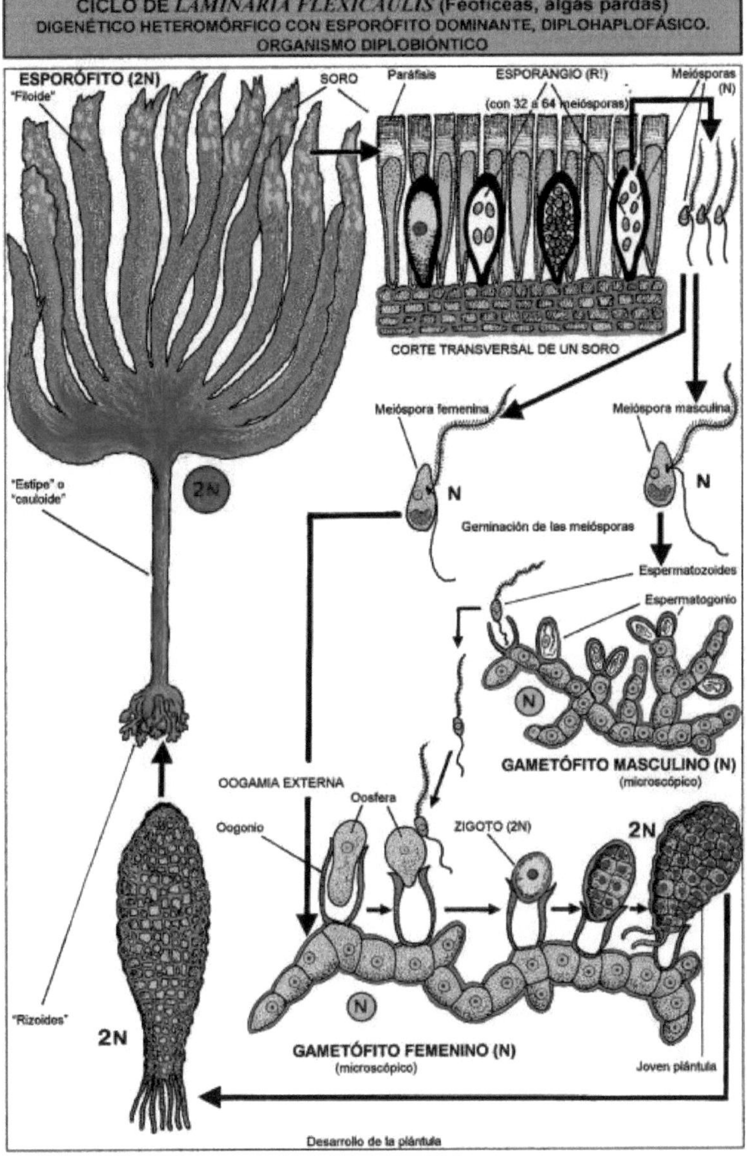

Figure 11: Sexual reproduction in *Laminaria flexicaulis* (Phaeophyceae)

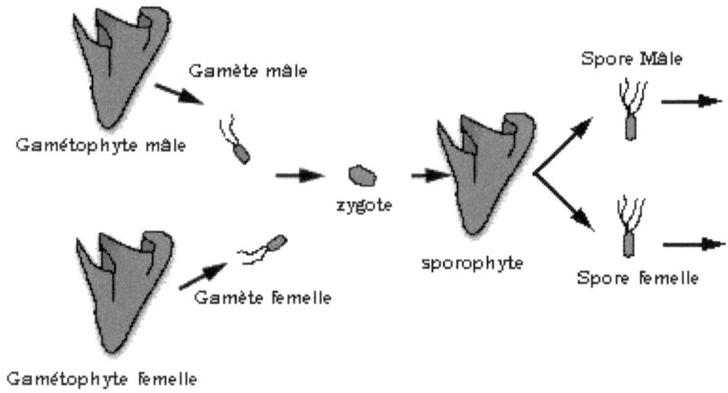

Figure 12: Algae life cycle showing the alternation between Gametophyte and Sporophyte phases.

13. Classification of algae :

Algae form a diverse group of aquatic plants, classified mainly according to their pigments, cell structure and mode of reproduction. Understanding their classification enables us to better appreciate their diversity and their role in marine and freshwater ecosystems. Here's an overview of the main algal lineages.

- **Chlorophytes (green algae) :**
  - ✓ **Chlorophyceae**: Contain chlorophyll a and b pigments, as well as carotenoids.
  - ✓ **Ulvophyceae**: Includes orders such as Bryopsidales and Cladophorales.
  - ✓ **Prasinophyceae:** a representative example is *Prasinochloris*.
  - ✓ **Streptophytes:** Includes several groups, including Mesostigmatophyceae, Zygnematophyceae, and Charophyceae.

- **Chromophytes (colored algae) :**

- ✓ **Bacillariophyceae (Diatoms):** Contain chlorophyll a and c pigments, as well as fucoxanthin.
- ✓ **Xanthophyceae (Yellow Algae):** A notable example is *Vaucheria*.
- ✓ **Fucophyceae (Brown Algae):** Includes orders such as Fucales and Laminariales.

➢ **<u>Rhodophytes (Red Algae) :</u>**

- ✓ **Rhodophyceae**: Possess chlorophyll a and d pigments, as well as bilichromoproteins such as phycoerythrin and phycocyanin. Orders include Bangiales and Gigartinales.

➢ **<u>Euglenophytes :</u>**

- ✓ **Euglenophyceae:** This group includes unicellular freshwater algae, some of which are photosynthetic with chlorophyll a and b.

➢ **<u>Dinophytes (Dinoflagellates) :</u>**

- ✓ **Dinophyceae:** A group of unicellular planktonic algae, some of which are toxic and responsible for red water phenomena.

➢ **<u>Haptophytes :</u>**

- ✓ **Prymnesiophyceae (Haptophytes) :** Unicellular, biflagellate algae with a coccolith exoskeleton.

➢ **<u>Cryptophytes :</u>**

- ✓ **Cryptophyceae**: Single-celled marine or freshwater algae, carrying chlorophyll a and c pigments, and mobile by flagella.

➢ **<u>Glaucophytes :</u>**

- ✓ **Glaucophyceae:** Includes genera such as *Glaucocystis* and *Cyanophora*, with an ancestral chloroplast called cyanelle, and ancestral characteristics.

This classification highlights the diversity of algae and their adaptation to various aquatic environments, both freshwater and marine.

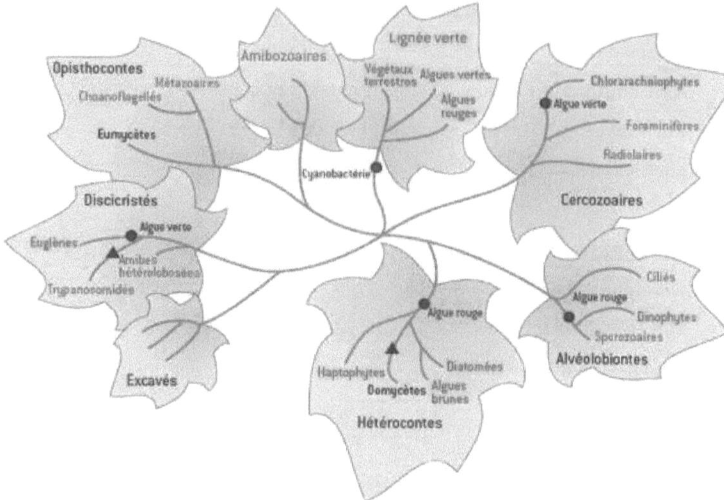

Figure 13: Classification of algae

14. Algae determination :

With the exception of a few common species, the identification of algae is a complex exercise. This task is particularly difficult for red algae, where many species resemble each other. Precise identification often requires consultation of determination keys, which vary in quality and effectiveness. To facilitate this task, it is useful to have recourse to complementary resources such as on-line photographs and specialized guides.

In mainland France, the majority of the 8,000 species of macrophytes have exclusively Latin names, which testifies to the general lack of interest in these plants.

It should be noted that no macroalgae harvested alive, i.e. attached to a substrate, is toxic. However, wrecked or floating algae should be avoided for food use.

15. Food use of seaweed :

Seaweed, often referred to as a veritable "vegetable of the sea", is composed of between 70% and 95% water, and is distinguished by its richness in minerals, surpassing land plants in calcium, magnesium and iron. They also contain trace elements and vitamins such as vitamins A, C and E, as well as provitamin A in the case of red algae, and vitamin B12, which is rare in plant sources and found in algae such as *Porphyra umbilicalis* and *Chlorella*. However, their nutritional contribution can vary according to their digestibility and assimilation. What's more, algae are rich in phlorotanins, fiber and, unlike terrestrial plants, offer complete proteins, providing all the essential amino acids.
Edible algae can be classified into several groups: green algae (such as sea lettuce, ao nori and ulva), red algae (such as dulse, nori and carrageenans), brown algae (such as wakame, kombu, arame, sea beans and sea spaghetti), and blue algae (mainly spirulina). They can be harvested in both coastal and marine environments, and can be macroalgae or microalgae.

Figure 14: Main food macroalgae and microalgae (spirulina, dulse, sea lettuce, Atlantic wakame, sea beans, kombu, wakame, nori and chlorella).

Table 1: Average composition of fresh Breton seaweed

Composition (per 100g)	Royal kombu (*Saccharina latissima*)	Wakame (*Undaria pinnatifida*)	Atlantic wakame (*Alaria*	Sea bean (*Himanthalia elongata*)	Sea lettuce (*Ulva* sp.)	Dulse (*Palmaria palmata*)

			esculenta)			
Energy (kcal)	35,0	18,0	26,0	23,0	22,0	24,5
Water (g)	86,2	90,9	89,4	88,2	88,6	88,4
Protein (g)	1,8	1,7	2,0	1,3	1,6	2,1
Fat (g)	0,3	0,2	0,3	0,3	0,3	0,2
Carbohydrates (g)	3,8	0,5	1,0	1,7	1,1	1,3
Fiber (g)	5,4	3,6	5,6	4,6	4,2	4,5
Salt (g)	1,3	1,8	2,1	1,3	2,2	1,3
Potassium (mg)	962,1	813,4	699,4	1181,4	258,5	1315,9
Magnesium (mg)	113,1	112,5	164,9	110,4	340,3	59,6
Calcium (mg)	140,6	66,5	124,1	95,1	95,1	105,4
Iodine (mg)	88,3	1,3	2,1	0,8	0,7	1,1
Iron (mg)	2,9	0,3	1,5	0,2	2,6	2,4
Beta-carotene (mg)	0,8	0,5	1,6	0,1	0,7	0,5
Vitamin C (mg)	12,7	6,5	19,4	36,8	8,0	7,2
Vitamin B9 (µg)	33,4	40,1	76,3	16,3	22,3	68,6
Vitamin K1 (µg)	44,4	66,8	30,0	5,9	2,1	73,1
Polyphenols (mg)	145,0	152,0	408,0	723,0	44,0	70,0

Fucoxanthin (mg)	7,8	6,8	14,4	4,9	-	-

Table 2: Contribution of 7 g dried seaweed to recommended daily intake

	RDA	Royal bleached kombu		Wakame		Atlantic Wakame		Sea beans		Sea Lettuce	
		Content	% RDA/7 g	Content	% RDA/7 g	Content	% RDA/7 g	Content	% RDA/7 g	Content	% RDA/7 g
Potassium (mg)	2000	81,5	4,1	611,3	30,6	337,1	16,9	577,8	28,9	167,8	8,4
Magnesium (mg)	375	84,7	22,6	63,7	17,0	67,6	18,0	52,8	14,1	140,7	37,5
Calcium (mg)	800	110,6	13,8	57,5	7,2	76,0	9,5	52,3	6,5	64,6	8,1
Iodine (mg)	0,15	3,7	2467	2,6	1733	2,7	1800	0,5	333	1,2	800
Iron (mg)	14	0,9	6,4	0,8	5,4	2,1	15,0	0,1	0,8	1,0	7,2
Manganes	2	0,0	0	0,0	2,5	0,1	6	0,2	7,5	0,1	6,5

e (mg)										
Provitamin A (µg)	800	70,4	8,8	1,1	0,1	11,7	1,5	37,3	4,7	11,7	1,5
Vitamin C (mg)	80	0,1	0,1	0,7	0,9	0,0	0	3,2	4	0,1	0,1
Vitamin B9 (µg)	200	6,4	3,2	17,4	8,7	26,5	13,3	6,1	3,1	1,2	0,6
Vitamin B12 (µg)	2,5	0,2	8,0	0,2	6,3	0,1	3,9	0,1	2,8	0,3	12,0
Vitamin K1 (µg)	75	67,7	90,3	23,8	31,7	61,0	81,3	23,4	31,3	0,6	0,8
	RDA	**Dulse**		**Nori**		**Maerl**		**Spirulina**		**Chlorella**	
		Content	**% AJR/7g**	**Content**	**% AJR/7g**	**Content**		**Content**	**% AJR/7g**	**Content**	**% AJR/7g**
Potassium (mg)	2000	419,5	21,0	121,3	6,1	6,4	0,3	95,2	4,8	69,5	3,5
Magnesiu	375	14,8	3,9	34	9,1	186	49,6	39,2	10,5	25,1	6,7

m (mg)											
Calcium (mg)	800	17,9	2,2	22,3	2,8	2131	266,4	34,1	4,3	25,7	3,2
Iodine (mg)	0,15	0,1	67	0,4	267	0,5	333	0	0	0	0
Iron (mg)	14	1,5	10,5	2,6	18,6	10,1	72,1	5,6	40,0	14,2	101,4
Manganese (mg)	2	0,6	29,8	0,3	15,0	1,2	60,0	0,3	15,0	0,2	10,0
Provitamin A (µg)	800	18,1	2,3	326	40,8	-	-	2058	257,3	3150	393,8
Vitamin C (mg)	80	0,1	0,1	4	5	-	-	0,8	1	5,3	6,6
Vitamin B9 (µg)	200	15,8	7,9	1,5	0,8	-	-	4,2	2,1	107,2	53,6
Vitamin B12 (µg)	2,5	0,3	10,2	2,7	108	-	-	16,5	660	26,9	1076,0
Vitamin	75	30,0	40,0	-	-	53,2	70,9	93,8	125,1	53,2	70,9

K1 (µg)											
RDA: Recommended Daily Allowance											

CHAPTER 2: MICROALGAE

1. Introduction :

The main components of phytoplankton, microalgae, including cyanobacteria, are photosynthetic single-cell organisms that have inhabited oceans and rivers for over three and a half billion years. The use of algae dates back thousands of years. Scientists have discovered that phytoplankton has been consumed in Mexico since the time of the Aztecs, while Chadians have been eating dried spirulina for several decades.

In Europe, it was against a backdrop of food shortages that researchers began to take an interest in microscopic algae as a food or dietary supplement in the 1940s. Their high protein content offered a solution to malnutrition problems. The first industrial plant dedicated to chlorella cultivation, initially developed to feed prey to juvenile farmed fish, was built in Japan in the 1960s.

In France, the first uses of microalgae date back to the late 1970s, with the appearance of the first mollusc hatcheries and the production of feed algae for aquaculture. Over time, companies specializing in microalgae production, known as microalgae growers, developed their production processes and, above all, added value to their production by marketing extracts. New applications have emerged in the nutraceutical food industry, as well as in the cosmetics industry, particularly in the field of cosmeceuticals. Other players in France have specialized in the manufacture of reagents for diagnostics and control. For example, two French companies, Alpha Biotech and Greensea, produce and sell allophycocyanins on the international market. Between 1975 and 2000, worldwide production of microalgae grew exponentially, from less than 5 tonnes to 3,500 tonnes. In 2004, worldwide production of microalgae, all species combined, was estimated at between 7,000 and 10,000 tonnes of dry matter, with a global market value of over 4.5 billion US dollars. At that time, 276 companies were listed worldwide in this field, a third of them producing essentially the three dominant species: *Spirulina*, *Chlorella* and *Dunaliella*.

Today, despite the cultivation of around ten species of micro-algae, global production remains stable at around 10,000 tonnes per year, a negligible figure compared to the 15 million tonnes of macro-algae produced annually. The most widely cultivated microalgae species are, in descending order: the cyanobacterium *Arthrospira* (spirulina), which accounts for around 50% of global production, followed by the green microalgae *Chlorella, Dunaliella, Haematococcus, Nannochloropsis*, and the diatom *Odontella*.

Asia dominates global microalgae production, accounting for around 50% of the total. Other major producing countries include the USA, Chile, Argentina, Israel and Australia. In Europe, Germany and the Netherlands are the main producers, with around 50 tonnes per year. France, meanwhile, began developing its first microalgae production units in the late 1980s. Today, there are some thirty production sites in France, with an annual output of around 10 to 15 tonnes.

Despite these advances, with the exception of a few species, microalgae have not yet reached a sufficient level of industrial maturity. Problems relating to quality consistency and production costs persist, limiting their access to certain markets.

2. Characteristics of microalgae :

2.1 Diversity of families and species :

Microalgae, found in both fresh and marine waters, are far more diverse than all land plants. It is estimated that there are at least 200,000 different species worldwide, and some authors put the figure at over a million species. These organisms form an extremely diverse polyphyletic group, comprising both prokaryotes (such as blue-green algae or cyanobacteria) and eukaryotes (including green, red and brown algae). Their classification into divisions is based on various properties, such as pigmentation, the chemical nature of the storage products of photosynthesis, the organization of photosynthetic membranes, and other morphological characteristics.

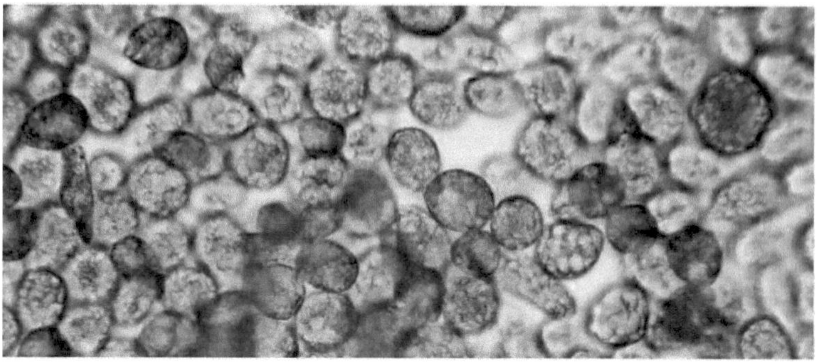

Figure 15: Microalgae.

Red and green algae belong to the same family, the Plantae, and are thought to have appeared around two billion years ago. Brown algae and diatoms, on the other hand, are thought to derive from a symbiotic association between a red microalga and a heterotrophic unicellular organism that emerged a billion years ago. The latter form the Chromalveolate family. It should be noted that some scientific studies tend to bring red algae closer to brown algae, taking into account different biogenetic markers.

Cyanobacteria, also known as blue-green algae, are prokaryotes whose most commonly cultivated species is spirulina. They appeared around 3.8 billion years ago, and are thought to have been responsible for the production of oxygen in the primitive atmosphere through photosynthesis. Their cells have a prokaryotic structure typical of bacteria, with photosynthesis taking place directly in the cytoplasm. Cyanobacteria are thought to be the origin of eukaryotic cell chloroplasts, enabling plants to photosynthesize through a process of endosymbiosis.

In eukaryotic microalgae, photosynthesis takes place in specialized structures, surrounded by a double plastid membrane, called chloroplasts. These cellular organelles, which contain DNA, are similar to those of cyanobacteria, which validates the hypothesis of endosymbiosis at the origin of these structures.

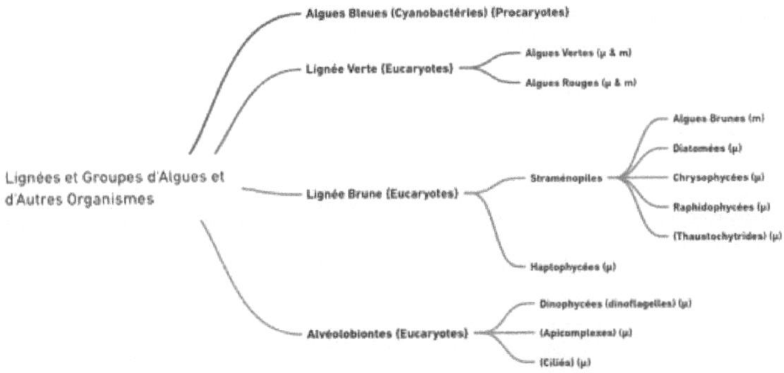

Figure 16: Phylogeny of algae from "The genetic classification of living organisms".

As far as French microalgae collections are concerned, the Roscoff Culture Collection (RCC) at the Roscoff Biological Station is particularly noteworthy. This international collection, created in 1998, aims to conserve micro-algae strains over the long term and make them accessible to fundamental and applied research, as well as to teaching in France and beyond. The CCR is also dedicated to the isolation of new strains, while guaranteeing the quality and traceability of the biological resources it holds.

In 2011, the CCR housed over 1,900 strains of microalgae and cyanobacteria, representing more than twenty classes of algae and protists from a wide variety of environments, including the English Channel, North Atlantic, North Sea, Mediterranean, Red Sea, Arctic Ocean, equatorial Pacific, tropical Atlantic and Indian Ocean. Recently, the collection has included photoheterotrophic bacteria and viruses from micro-algae. The CCR is one of the five largest collections of marine microalgae cultures in the world, providing an essential biological resource for exploring the taxonomic diversity of phytoplankton and enabling a wide range of fundamental and applied research.

In France, two other noteworthy microalgae collections are the Algobank, located in Caen, which includes almost 400 species, and the Institut Pasteur culture collection, specializing in cyanobacteria.

In addition, several laboratories such as CNRS/UPMC LOV, IFREMER PBA, and CEA/CNRS LB3M have developed their own strain libraries to manage the various species and strains they work on.

2. 2 Biology, lifestyle and reproduction :

Some species of microalgae have the ability to switch from photo-autotrophic growth, where they use light to convert CO_2 into carbon chains, to heterotrophic growth, which takes place without light and uses glucose or other carbon substrates for their carbon and energy metabolism. Some algae can also adopt mixotrophic growth, combining the two modes.

Microalgae have the advantage of a very short division cycle, often of the order of a few hours, enabling rapid biomass production, reaching several grams of dry matter per liter. Their development depends on a number of growth factors and culture conditions, such as water, nutrients, light, CO_2, temperature, culture pH and agitation. Depending on the strains cultivated, the water used can be fresh, marine or brackish.

Water supply and water quality are crucial factors in determining and influencing algae cultivation. To ensure optimal growth in photobioreactors, it is often necessary to start cultivation in sterile water, devoid of any other micro-organisms or molecules that might inhibit or compete with algal growth, or at least to seed it with a significant quantity of biomass.

The nutrients required for algal growth vary according to trophic mode, cultivated strain and water source. In autotrophic mode, microalgae can use nitrogenous (nitrate, nitrite, ammonium) and phosphatic (phosphate) mineral forms. Whatever their growth mode, algae also require potassium, iron, silica (for diatoms), sulfur, trace metals and vitamins.

Interestingly, some nutrient deficiencies are deliberately applied to stimulate the production of certain metabolites. For example, a nitrogen, phosphorus or silica deficiency can induce, in certain species, a high accumulation of lipids. In heterotrophic mode, an organic carbon source is used, such as sugars, organic acids or glycerol.

As with all chlorophyllous plants, photosynthesis enables the fixation of carbon dioxide in the atmosphere or dissolved in water, using light energy to produce biomass. Algae use different chlorophyll pigments to capture photons of different wavelengths. Depending on the application and cultivation area, natural (sun) light can be used, or the algae can be illuminated by an artificial light source (neon, LED, etc.).

Temperature, pH, dissolved inorganic carbon and culture homogenization rate are also crucial factors for cultivation. Depending on the strain being cultivated, there are optimal ranges for these parameters that guarantee optimum growth. If conditions deviate from these ranges, culture performance can be significantly degraded.
Among the factors limiting growth is the accumulation of oxygen in the medium, produced by photosynthesis in a closed system. It is also important to monitor the risk of contamination by other micro-organisms, whose rapid growth could overtake that of the desired strains, or by predators likely to rapidly consume the biomass. Viruses can also cause malfunctions in cultivation processes. However, this problem is largely solved for algae growing in extremophilic environments, such as hypersaline waters (like *Dunaliella salina*) or hyperalkaline waters (like spirulina), which limit the growth of predators and competing micro-organisms.

2.3 Biochemical composition :

Microalgae are distinguished by the great diversity of molecules present within their cells, differentiating them from other plants in their richness in lipids, proteins, vitamins, pigments and antioxidants. This biomass is a significant source of almost all essential vitamins, such as B1, B6, B12, C, E and K1, while possessing a wide range of pigments, both fluorescent and non-fluorescent, that can also act as antioxidants.
In addition to chlorophyll, which accounts for between 0.5 and 1% of dry matter and is the primary photosynthetic pigment in all photosynthetic algae, microalgae contain a variety of additional pigments, such as carotenoids (0.1 to 0.2% of dry matter) and phycobiliproteins (phycoerythrin and phycocyanin).
Among the most widely exploited pigments are phycocyanin from spirulina (blue colorant), phycoerythrin from *Porphyridium purpureum* (red color), astaxanthin from *Haematococcus pluvialis* and beta-carotene from *Dunaliella salina*.
Microalgae can accumulate over 50% of their dry weight in lipids, mainly in the form of triglycerides, phospholipids and glycolipids. These lipids contain saturated and polyunsaturated fatty acids (PUFAs), such as omega-3 (ALA, EPA, DHA) and omega-6 (ARA). The high protein, peptide and amino acid content (between 12 and 65% of dry matter) of many microalgae species makes them an unconventional source of

protein, particularly interesting for human and animal nutrition, especially in fish farming.

Some species of microalgae are also rich in oligosaccharides and polysaccharides, while others can produce molecules with antiviral, antibiotic or antiproliferative activity in humans. It is likely that many molecules have yet to be discovered in microalgae, and their potential is the subject of intensive research in many laboratories around the world.

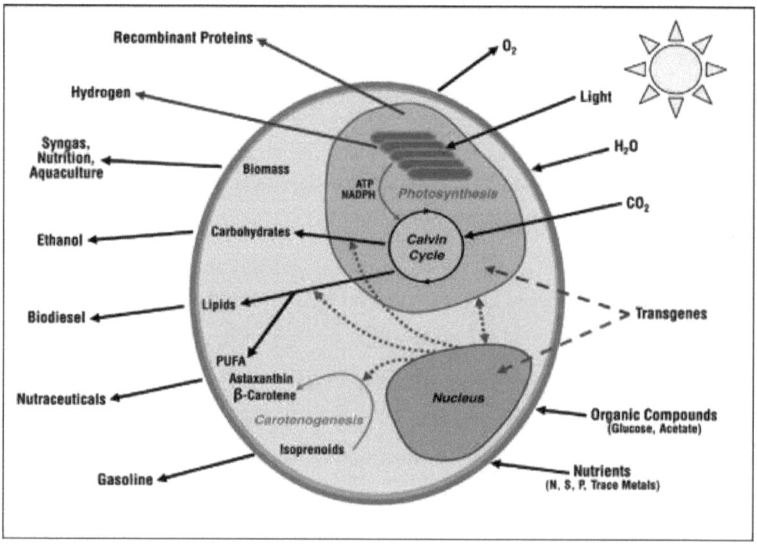

Figure 17: Microalgae operating diagram, inputs, products, applications.

3. Biomass production :

The production of algal biomass has evolved considerably over time. In France, the first algae producers appeared with the first hatcheries in the late 1970s. At the time, microalgae were mainly grown outdoors, often in batch ponds under semi-controlled conditions. Since then, production techniques have improved, enabling continuous, controlled production in photobioreactors, while open systems have also benefited from optimization and automation.

In some cases, microalgae and macroalgae are grown together under symbiotic conditions, benefiting both types of production.

For example, *Odontella* and *Chondrus* are co-cultivated in the same flow-through tanks at Innovalg.

Microalgae and cyanobacteria can be grown photo-autotrophically, in open or closed systems of various sizes and geometries, using sunlight and/or artificial light. They can also be cultivated heterotrophically, a well-known and long-established method for growing bacteria. Each of these cultivation methods is presented below.

4. In photoautotrophic cultures

4.1 Open systems: natural ecosystems, lagoons and high-yield ponds (Raceways).

Open systems are the most commonly used for commercial microalgae production, due to their ease of use, relatively low cost and standardization. Among these systems, high-yield ponds, also known as Raceways, are particularly widespread. These ponds are shallow, with a maximum depth of 30 cm, and can be circular or elongated in shape, alone or connected to each other.

The water in these ponds is generally kept in motion by paddlewheels or rotating structures, enabling efficient mixing thanks to well-designed guides. Raceways remain the most widely used technology in this context.

The nutrients required for microalgae growth in these systems are generally supplied by inorganic agricultural fertilizers. However, they can also be supplied by run-off from nearby agricultural areas, or by channelling water from wastewater or water treatment. Algae cultures in these ponds can be defined, with one or more selected strains, or be made up of an indefinite mixture of strains.

Figure 18: Raceways open production system.

Open ponds offer a number of major advantages, including easy, low-cost construction and rapid commissioning. However, these systems, which require little initial investment, offer limited control over the crops, resulting in significantly lower productivity than photobioreactors.

Several factors contribute to this limitation:

- ✓ **High pond depth (20-50 cm)**: Light penetration is limited, resulting in low biomass concentration and reduced volumetric productivity.
- ✓ **$CO_{(2)}$ supply**: In these systems, carbon dioxide supply depends on either atmospheric concentration or bubbling. However, the transfer of CO_2 into the liquid phase is often inefficient due to the shallow depth of the basins.
- ✓ **Dependence on natural conditions**: Uncontrolled variations in temperature, wind and light intensity in a natural environment lead to random, seasonal productivity.
- ✓ **Vulnerability to contamination**: The openness of the system makes crops susceptible to contamination by external parasites or predators, compromising

the stability of productivity over long periods. However, this type of system is well suited to the cultivation of extremophilic microalgae such as *Dunaliella salina* and *Spirulina*, which thrive in highly saline or alkaline environments respectively.

- ✓ **Lack of agitation**: The absence of sufficient agitation in the culture volume reduces the algae's continuous exposure to light, thus limiting their growth.
- ✓ **Water requirements**: An additional disadvantage is the need for high water inputs to compensate for the high evaporation in these cropping systems.

These factors explain why, despite their advantages in terms of cost and simplicity, open ponds cannot achieve the same levels of productivity as more sophisticated systems such as photobioreactors (PBRs).

4.2 Closed PBR systems :

Photobioreactors (PBRs) are closed tanks or systems designed for the cultivation of algae, particularly suited to microalgae sensitive to contamination, as well as to extremophilic algae. These systems enable the cultivation of one or more specific strains, optimized for the production of the desired product.

PBRs can be operated in continuous or discontinuous mode, with continuous cultivation generally offering higher biomass productivity.

Figure 19: Tubular reactor.

Water, the necessary nutrients and CO_2 are supplied in a controlled manner in these systems, with optimized transfer of CO_2 from the gas phase to the liquid phase. Due to the high productivity of these systems, it is also necessary to continuously remove oxygen to avoid growth inhibition.

Depending on production objectives and climatic conditions, closed PBR systems can be exposed to artificial light or outdoor sunlight. Thermal control systems may be required to maintain the crop at a constant temperature, as sunlight can cause the crop to warm up, without being compensated for by water evaporation.

Unlike open systems, photobioreactors are more expensive to build and more complex to set up, requiring more stringent maintenance and regular cleaning procedures.

However, they offer better crop control, more stable and sustainable production over time, and consequently significantly higher yields.

There are several types of photobioreactor, adapted to the biology of the algae and their cultivation environment. Here are the three main types:

- ✓ **Vertical column airlift photobioreactors**: These reactors are simple in design, but the large diameters (around ten centimetres) often used result in average volumetric efficiency. A bubbler at the base of the reactor provides the necessary mixing.

- ✓ **Tubular photobioreactors**: These reactors are an alternative to column reactors. They can form a network of horizontal, vertical, oblique, conical or serpentine tubes. Small tube diameters (a few centimeters) enable high productivity. However, pressure drops in the tube network can be significant, and for long tubes, pH, CO_2 and dissolved O_2 gradients, as well as fouling phenomena, can occur, affecting system productivity.
- ✓ **Flat photobioreactors**: Introduced by Milner in 1953, these reactors offer a large illuminated surface, with very low culture thicknesses of a few centimetres, or even less than a centimetre (such as the AlgoFilm photobioreactor developed as part of the ANR BIOSOLIS program). They can be installed outdoors, and have a lower dissolved oxygen build-up than horizontal tubular photobioreactors. However, they are also prone to fouling.

Figure 20: BioFence tubular reactor.

To solve the problem of access to light in photobioreactor cultures, systems illuminated inside the culture volume by fluorescent lamps or optical fibers, possibly coupled with solar capture systems (such as the DiCoFluV photobioreactor developed in the ANR BIOSOLIS program), have been proposed. These systems can achieve very high productivity, but the technology, as well as cleaning and maintenance procedures, are more complex.

There are also projects for photobioreactors immersed in ponds or in marine environments. Immersion provides both mechanical support for the tubes (generally flexible sleeves, possibly micro- or nano-porous) and thermal equilibrium for their contents, buffered by the surrounding water.

Despite their many advantages, photobioreactors also have some disadvantages, including:

- ✓ **Oxygen accumulation at inhibitory levels**: the formation of free radicals due to oxygen accumulation can limit productivity.
- ✓ **Significant overheating at solar maximum**: the greenhouse effect can require large quantities of water to mist the air and keep the temperature outside the zones that are lethal to cells.
- ✓ **Pressure losses linked to fluid circulation**: These are often significant, resulting in high energy consumption, which can be a disadvantage when it comes to producing bioenergy.
- ✓ **Difficulty of cleaning in the event of contamination**: In the event of contamination, these systems can be extremely tricky to clean to remove the contaminant.

5. In heterotrophic culture :

Microalgae can be cultivated using conventional biotechnological processes, using heterotrophic algae or making them heterotrophic, and producing them in fermenters rather than photobioreactors. In this type of bioreactor, light energy is not used as an energy source, thus eliminating the problem of homogeneous light diffusion. The necessary energy is supplied by a carbon substrate, which means that algae do not photosynthesize, but breathe.

The advantages of this production system are numerous:

- ✓ **High volumetric productivity**: Volumetric productivity is considerably increased, up to 100 times that of an open system, and 10 times that of a photobioreactor.
- ✓ **High dry matter concentrations**: Dry matter concentrations can reach several hundred grams per liter.
- ✓ **Low production costs**: The system offers relatively low production costs.
- ✓ **Product quality**: The products we produce are of the highest quality.
- ✓ **Contained environment**: The contained system eliminates the risk of contamination.
- ✓ **Location flexibility**: There are no geographic location constraints.
- ✓ **Ease of industrialization**: The system is highly industrializable.
- ✓ **Mastered technology**: This technology has been fully mastered on an industrial scale, notably for the production of yeast and bacteria in industries such as chemicals and food, for several decades.

However, this system also presents certain constraints:

- ✓ **Necessity of adapted strains**: It is crucial to have strains adapted to this method of cultivation.
- ✓ **Installation and investment costs**: Initial reactor and installation costs are high.
- ✓ **Dependence on external carbon sources**: Cultivation requires external carbon sources, which must be sterilized and supplied in sufficient quantities to match the products generated.
- ✓ **Not suitable for bioenergy production**: This technology is not suitable for producing bioenergy from solar energy.
- ✓ **Inability to produce pigments**: The system is unable to produce pigments, which limits its application to certain specific productions.

As a result, this technology is mainly used for the production of high-value-added products, such as organic molecules, rather than for large-scale applications such as bioenergy production.

6. An overview of some major international installations :

- ✓ **Cellana (Hawaii, USA)**: The Cellana consortium operates a site comprising almost 2.5 hectares of Raceways in Hawaii, dedicated to the cultivation of micro-algae.
- ✓ **SD-CAB (San Diego-Center for Algae Biofuels, California, USA)**: This center is developing open-air Raceways on a 16-hectare site, under low-automation conditions. In parallel, more instrumented and innovative platforms are also being developed, albeit on a smaller scale.
- ✓ **Synthetic Genomics-EMRE (Exxon Mobil, California, USA)**: The Synthetic Genomics-EMRE consortium in California is the largest and probably the most automated greenhouse pilot to date. This greenhouse houses various types of photobioreactors and some fifteen Raceways.
- ✓ **AlgaeParc (Netherlands)**: Located at Wageningen University, AlgaeParc is a project comprising four modules of 25 m^2 each, combining various photobioreactor and/or Raceway technologies.
- ✓ **SALINALGUE (France)**: In the south of France, the SALINALGUE project brings together a large consortium of industrialists and laboratories led by La Compagnie du Vent. The project involves large-scale cultivation of a consortium of algae dominated *by Dunaliella salina* in basins converted from seaside salt works. The extremelyophilic environment created by the hypersaline water limits contamination and predation, providing a favorable environment for microalgae cultivation.

Figure 21: San Francisco Bay salt marshes.

7. Comparison of Micro-Algale biomass production systems: productivity and production costs

7.1 Production systems

Microalgae can be grown in a variety of systems, each with its own advantages and disadvantages in terms of productivity and production costs. The most commonly used systems are :

- ✓ **Photobioreactors (PBR)**
- ✓ **Raceways ponds**
- ✓ **Hybrid systems (combination of PBR and open basins).**

7.2 Productivity

Productivity in terms of microalgal biomass is influenced by several factors, including light, CO_2, temperature, algal strain, and nutrient quality. Here's a comparison between the systems:

- ✓ **Photobioreactors (PBR)**
  - ➢ **Advantages :**
    - High productivity thanks to better control of growing conditions.
    - Allows dense cultivation with low risk of contamination.
    - Optimizing light exposure.
  - ➢ **Disadvantages :**
    - High installation costs.
    - High energy requirements for mixing and circulating fluids.
  - ➢ **Productivity:** Generally higher than open ponds, with yields in the range of 1 to 5 g/L/day.

- ✓ **Open basins**
  - ➢ **Advantages :**
    - Relatively low installation and running costs.
    - Easy to implement for large-scale production.
  - ➢ **Disadvantages :**
    - Less control over environmental conditions, which can affect productivity.
    - Increased risk of contamination by other micro-organisms.
  - ➢ **Productivity:** Lower than PBR, with typical yields of 0.1 to 0.5 g/L/day.

- ✓ **Hybrid systems**
  - ➢ **Benefits:** Combines the advantages of both systems, with better control than open basins and lower costs than PBRs.
  - ➢ **Productivity:** Intermediate between PBR and open ponds.

7.3 Production costs :

The cost of producing microalgal biomass is influenced by several factors, including investment costs, operational costs (energy, labor, nutrients, CO_2), and the efficiency of biomass harvesting and processing. Here's a comparison:

- ➢ **Photobioreactors (PBR) :**
  - **High costs:** PBRs are expensive to install and operate due to infrastructure and energy requirements. Costs can vary from €10 to €20/kg of dry biomass, depending on the scale and efficiency of the system.
  - **Profitability:** Viable for high-value-added products (such as nutraceuticals, pigments, pharmaceuticals).

- ➢ **Open basins**

- **Low costs:** Open ponds are much cheaper to build and maintain, with production costs ranging from €1 to €5/kg of dry biomass.
- **Profitability:** Suitable for low value-added products, such as large-scale production of biofuels or dietary supplements.

➢ **Hybrid systems**

- **Intermediate costs:** Hybrid systems reduce costs by combining the advantages of PBRs for critical growth phases and open ponds for mass production. Costs can vary between €5 and €10/kg of dry biomass.

In summary, PBRs have the highest level of productivity in terms of micro-algal biomass, but are the most expensive. Open ponds have low productivity but are very inexpensive. Hybrid systems aim to be a compromise between these two extremes, maximizing both productivity and economy. In terms of manufacturing costs, open basins are more efficient for larger plants, while PBRs are recommended for specific, high value-added products. Consequently, the type of production structure to be used will depend to a large extent on the requirements, both in terms of volume and quality, and the economic limitations of the project.

8. Harvesting and concentration processes

8.1 Introduction

The microalgae harvesting and concentration stage is now recognized as a critical point in the autotrophic industry, representing up to 20-30% of algal biomass production costs. This stage is influenced upstream by the choice of species and cultivation method, while also conditioning downstream stages such as extraction, purification and recycling.

The diversity of microalgae strains, with small cell sizes ranging from 1 to 100 micrometers, and highly diluted suspensions from 0.1 to 10 grams per liter, further complicates this stage. These parameters must be carefully considered when selecting the harvesting technology.

There are also special cases of production where the micro-algal biomass is not physically harvested, but only the substances of interest produced by the cells are recovered, before restarting a culture cycle. This process is known as milking. This production method has been used in an exemplary way for lipid production. In this case, the algae are brought into contact with an organic extraction solvent, and a liquid/liquid extraction operation is carried out, followed by phase separation. The treated algae are then re-cultured, with survival rates reaching close to 100% in some cases.

Possible harvesting methods are based on two distinct separation principles:

- ✓ **Principle of separation by density difference**: Techniques based on this principle include sedimentation, flocculation-settling, flotation and centrifugation.
- ✓ **Size exclusion separation principle**: This principle includes sieving and membrane processes such as microfiltration and ultrafiltration.

8.2 Gravity sedimentation by density difference :

Some strains of microalgae have a natural tendency to settle when all forms of medium mixing are stopped. This property facilitates their harvesting, as it enables the algae to be concentrated, thus reducing the volume to be processed. The settling process depends on three main parameters:

- ✓ **The difference in density** between the culture medium and the micro-algae: A greater difference in density accelerates sedimentation.
- ✓ **Microalgae size**: Larger microalgae tend to settle faster.
- ✓ **The viscosity of the culture medium**: According to Stokes' law, the viscosity of the medium influences the sedimentation rate, with higher viscosity slowing down the process.

By combining these factors, it is possible to optimize settling time and improve the efficiency of microalgae harvesting by gravity sedimentation.

Figure 22: Different sedimentation times.

8.3. Flocculation-settling :

The phenomenon of flocculation involves the aggregation of micro-algae cells, facilitating their decantation. Several flocculation methods are used, including chemical flocculation and bioflocculation:

- ✓ **Chemical flocculation**: This process uses chemical additives to bind the algae or modify the physico-chemical interactions between them. Examples include the addition of iron salts ($FeCl_3$, $Fe2(SO_{4)(3)})$ or aluminum salts ($Al_2(SO_{4)(3))}$, or the use of synthetic or natural anionic or cationic polymers.

- ✓ **Bioflocculation**: This method can be triggered by a change in environmental conditions (pH, light, temperature, nutrient deficiency), resulting in cellular stress that generates the synthesis of aggregating exudates, a process known as autoflocculation. Bioflocculation can also be triggered by the use of microbial populations that produce aggregating exudates.

- ✓ **Electrocoagulation and electroflocculation**: These methods have the advantage of not using chemical additives. Electrocoagulation uses iron or aluminum electrodes through which a current flows, releasing metal ions by oxidation-reduction. Electroflocculation, on the other hand, can use electrodes or ultrasound. With electrodes, cells lose their charge when they come into contact with the anode (+), allowing them to agglomerate. Ultrasound generates acoustic fields, concentrating cells in areas of minimum potential, causing them to aggregate.

8.4 Flotation :

Some strains of microalgae, particularly those rich in lipids, have a natural tendency to float. For others, flotation is induced by acting on the difference in density and cell diameter. Two main methods are used:

- ✓ **Electroflotation**: This process enables algae to float to the surface thanks to the action of hydrogen microbubbles generated by water electrolysis. The result is a dense layer of algae suspended on the surface. However, this method has a high energy consumption and can pose problems for electrode durability in marine environments, due to the presence of chlorides.

- ✓ **Dissolved air flotation (DAF)**: Initially developed for wastewater treatment, this method involves bubbling air into the algal suspension, allowing the algae to float to the surface. Flocculation can be used to enhance the process. The effectiveness of this method depends on the size and distribution of the air bubbles in the suspension. The algae-enriched top layer is then recovered for further processing.

These methods recover algae concentrate with a dry matter content of 1-6%.

8.5 Centrifugation: action on the g number :

Centrifugation uses centrifugal force to separate suspended solid particles, such as algae, from the fluid in which they are suspended. This technique relies on gravitational intensity (g-number) to achieve separation, and the apparatus used is called a centrifuge. Several operating parameters influence the efficiency of this technique, including :

- ✓ **Cell characteristics** (size, concentration, etc.).
- ✓ **Dwell time** (controlled by feed rate).
- ✓ **Acceleration** (number of g).

Several types of centrifuge exist, such as the bowl centrifuge, the plate separator, the centrifugal decanter and the hydrocyclone. Although centrifugation is one of the most commonly used techniques, it is also among the most costly in terms of investment and energy consumption.

8.6 Face filtration: sieving, separation by size exclusion :

Frontal filtration involves passing the fluid to be filtered perpendicular to the filter surface. This technology is particularly suitable for large cells, in excess of 40 µm. This process is simple to implement, using cloths or sieves for harvesting certain algae such as spirulina.

Frontal filtration operates under low pressure gradients (or vacuum) and requires regular scraping of the algae cake formed on the filter to maintain filtration rates. However, this technique is limited by the accumulation of particles on the filter surface, which can lead to clogging.

Depending on the nature of the strain being cultivated, pre-concentration or a pre-filtering layer may be necessary. The dry matter content and energy consumption depend on the type of sieve used.

8.7 Crossflow membrane filtration: separation by size exclusion :

Tangential filtration, also known as tangential flow filtration, involves passing the fluid parallel to the filter surface, unlike frontal filtration. In this process, the pressure of the fluid allows the liquid to pass through the filter, while the particles remain in the tangential circulation flow; this considerably reduces the risk of clogging.

This technique is particularly well-suited to very small particles, less than 40 μm in size; this includes the majority of single-cell micro-algae. Thanks to this method, it is possible to maintain a more stable and prolonged filtration flow, while minimizing the accumulation of particles on the filter surface; a significant advantage for the treatment of small microalgae.

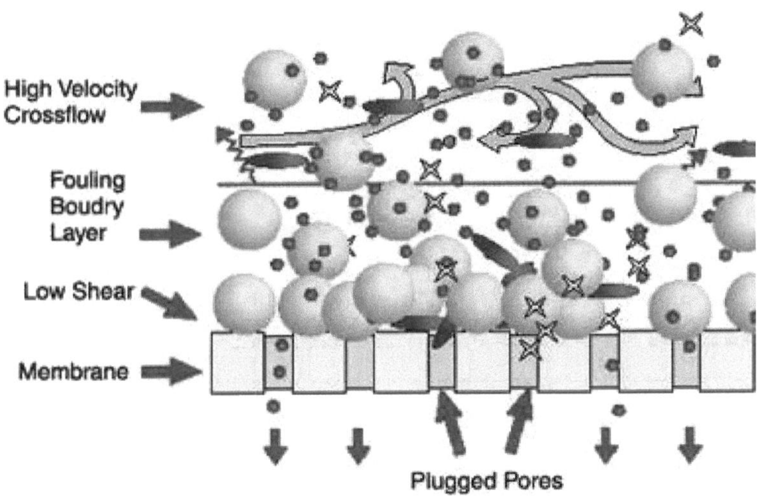

Figure 23: Principle of crossflow filtration.

In microalgae harvesting, good cell retention requires the use of microfiltration (MF) or ultrafiltration (UF) membranes, with pores ranging from 0.5 μm to 0.02 μm (≈ 50 kDa).

These systems, although efficient, are characterized by limited pressure difference and tangential velocity, as well as high energy consumption due to the pumping and pressure required to force the fluid through the membranes. Despite these challenges, they may represent a plausible alternative to centrifugation for microalgae harvesting. Similar systems, but equipped with mineral membranes, have been used for decades in the treatment of drinking water, in particular to make water that is resistant to conventional treatment potable, as is the case at the Méry-sur-Oise pumping station. Although tangential filtration is often the preferred mode for this type of filtration, experimental studies in frontal mode are currently underway at GEPEA, exploring new possibilities for optimizing these processes.

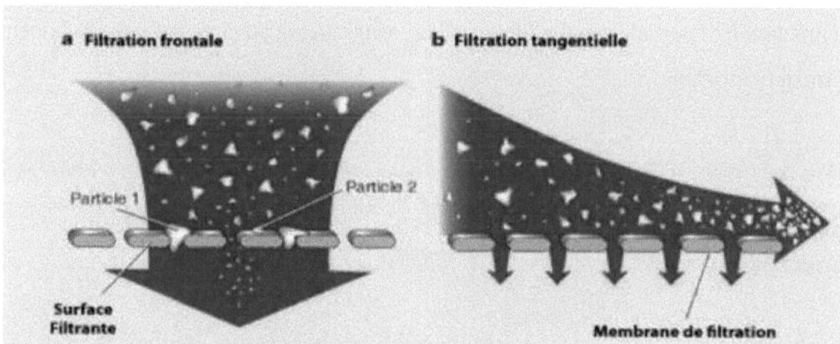

Figure 24: Comparison of the two filtration modes

8.8 Comparison of different harvesting systems :

It's important to note that the different microalgae harvesting techniques can be used in cascade, i.e. for initial harvesting followed by further concentration (harvesting & dewatering). This approach optimizes overall process efficiency by combining the specific advantages of each method.

Harvesting technologies are often used in combination to maximize efficiency and reduce costs.

For example:

- ✓ **Flocculation followed by flotation or decantation**: Flocculation aggregates the micro-algae, facilitating their separation by flotation or decantation.
- ✓ **Flocculation followed by centrifugation**: After flocculation, centrifugation can be used to further concentrate the algae, improving biomass recovery.

- ✓ **Flocculation followed by filtration**: Although less common, this combination allows flocculation to be used to improve filtration efficiency, reducing the risk of clogging and increasing yield.

These combinations are chosen according to the specific characteristics of the microalgae, the requirements of the production process, and the final objectives of the application, thus maximizing harvesting efficiency while minimizing costs and operational constraints.

8.8.1 By energy consumption :

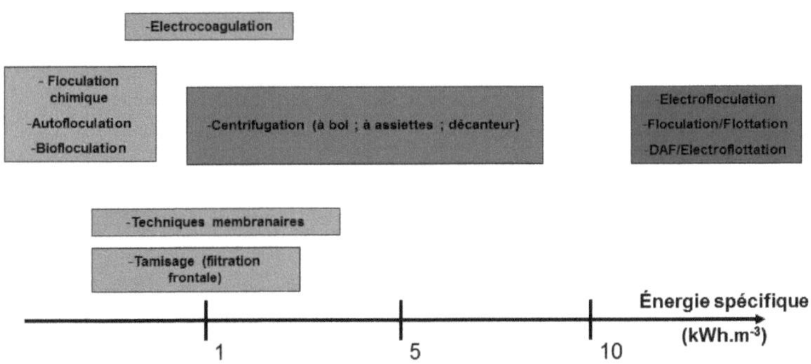

Figure 25: Comparison of harvesting systems according to energy consumption

8.8.2 Depending on the final biomass moisture target :

Table 3: Comparison of harvesting systems according to their final moisture target.

Very wet biomass (< 5% DM)	Wet biomass (5 - 30% DM)
- Natural sedimentation	- Sieving
- Flotation	- Filtration
- Chemical flocculation	- Centrifugation
- Bioflocculation	- Membrane techniques (micro or ultra)
- Electrocoagulation	

- Electroflocculation	

9. Conclusion :

There is no universal process for harvesting algae; the choice of method depends on several key parameters. The strain of microalgae used, as well as its morphological, dimensional and physiological properties, may dictate the selection of one harvesting system over another.

The choice of harvesting technology must also take into account the crop throughput to be processed, processing efficiency (retention rate, final dryness), impact on biomass quality and further processing, process stability, and potential performance drifts (such as media evolution, bacterial growth, cell debris, and exopolymer release). Other considerations include the material used (equipment material in relation to the medium) and associated costs (investment, energy, labor, reagent consumption).

It is also important to note that an efficient harvesting process on a small scale will not necessarily be technically and economically viable on a large or very large scale for a given strain.

The final concentration of the microalgae paste will depend on the harvesting method used, and will have a direct impact on the amount of energy consumed, as well as on downstream processing and extraction processes.

Generally speaking, all these stages are energy-intensive, which is a major challenge when it comes to optimizing large-scale microalgae harvesting processes.

10. Pretreatment and extraction processes :

The extraction of molecules of interest from algal cells is a crucial step in the valorization of microalgal biomass. This biomass, with its high moisture content, poses challenges for the preservation and extraction of the products to be valorized. For example, prior drying is often necessary to optimize extraction. As the molecules of interest are generally locked inside cells, it is essential to overcome the barriers of cell membranes. Grinding the biomass prior to extraction, using mechanical or non-mechanical methods, improves the efficiency of this stage.

10.1 Drying :

Drying is a crucial step in preparing micro-algal biomass for extraction. Several drying processes are available, each adapted to different applications:

- ✓ **Solar drying**: Traditional method in which harvested and concentrated seaweed is spread out in a greenhouse or in the open air to allow the water to evaporate through the action of the sun.

- ✓ **Spray drying**: A modern technique in which a liquid is transformed into a powder by passing through a stream of hot air. Droplets sprayed into a cylindrical enclosure evaporate, leaving a powder which is then separated from the air.

- ✓ **Lyophilization**: This process consists of removing water from a product by deep-freezing followed by vacuum evaporation, thus preserving the volume, appearance and properties of the treated product. However, this method is energy-intensive.

- ✓ **Détente Instantanée Contrôlée (DIC)**: Presented by *the Laboratoire d'Etudes des Phénomènes de Transferts et de l'Instantanéité* at the University of La Rochelle, this technology treats biomass at high temperature and high pressure over a short period of time. An abrupt drop in pressure causes instantaneous self-vaporization, resulting in rapid cooling and texturing of the material. DIC is used for drying, decontamination and lipid extraction.

10.2 Grinding :

Grinding is a key step in allowing solvents to access the molecules of interest locked inside algal cells. There are several grinding methods, which can be mechanical or non-mechanical:

- ✓ **Mechanical**: cell homogenization, ball milling, ultrasound, autoclaving.

- ✓ **Non-mechanical**: Freezing/thawing, use of organic solvents, osmotic shocks, acid-base reactions, enzymatic lysis.

- ✓ **Other methods**: Use of microwaves or sonication (ultrasound) to break down cell membranes.

In some cases, efficient grinding may suffice to eliminate the need for complex extraction, enabling direct purification by filtration.

10.3 Extraction :

Extraction is a key step in isolating biological compounds of interest. The choice of method depends on the molecules to be extracted, which vary according to the algal species and its state of growth. A high degree of concentration is often required prior to extraction, which may involve prior drying.

10.3.1 Extraction with organic solvents :

- ✓ **Solvents used**: Hexane, chloroform, ethanol, isopropanol, butanol, ketones, esters, vegetable oils.

- ✓ **Process**: Solvents are chosen according to their polarity and affinity with the target molecules. The presence of water can limit extraction efficiency, requiring prior drying.

- ✓ **Industrial applications**: Extraction of beta-carotene from *Dunaliella salina* in India, maceration of dry micro-algae to extract cosmetic active ingredients.

10.3.2 Extraction with subcritical water :

- ✓ **Principle**: use of water at temperatures just below the critical temperature (Tc=376°C, Pc=221 bar); this reduces the polarity of the water and increases the solubility of organic compounds.

- ✓ **Benefits**: Shorter extraction times, higher quality extracts, lower extracting agent costs, environmental compatibility.

- ✓ **Challenges**: Difficult to extrapolate to large scale, high energy requirements for heating the system.

10.3.3 Extraction with supercritical fluids :

- ✓ **Principle**: Use of fluids above their critical point to enhance their solvation power. This process can be used in batch or continuous mode.
- ✓ **Advantages**: Easy separation of solvents and products after extraction, with solvents returning to their gaseous state once pressure and temperature return to atmospheric conditions.

- ✓ **Applications** : Supercritical fluid extraction is commonly used to isolate high-value molecules of interest.

These different extraction methods make it possible to take advantage of the molecules present in microalgae, with each method having its own advantages and limitations depending on the specific production objectives.

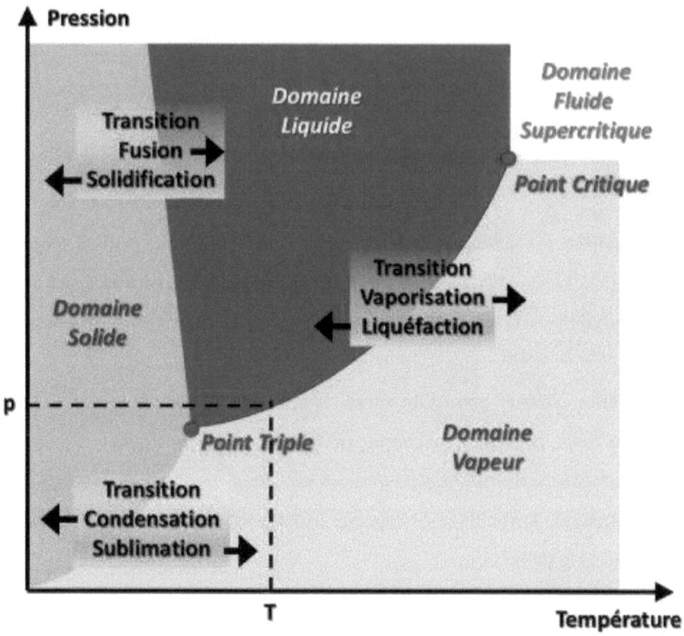

Figure 26: Diagram of the different states of a body.

Supercritical fluid extraction techniques are widely used for the commercial extraction of substances from solid substrates, such as the extraction of caffeine from coffee beans, an application which has been in use for over 20 years. The majority of applications use carbon dioxide (CO_2) as the supercritical fluid (Tc=31°C and Pc=74 bar). However, other solvents can also be used, such as ethane, hexane, water (Tc=376°C and Pc=221 bar), methanol, nitrous oxide, sulfur hexafluoride, as well as n-butane and pentane.

✓ **Note:**

Supercritical processing is not limited to extraction; it is also capable of simultaneously extracting and converting oils into biofuels. The technique has also proved effective in extracting other compounds from algae. The supercritical transesterification

approach can also be applied to algal oil extracts, offering a powerful method for producing biofuels directly from algal biomass.
Supercritical fluids have the advantage of being selective, enabling extracts of high purity and concentration to be obtained. What's more, this method leaves no residual organic solvents in the extract or in the treated biomass, which is a considerable advantage for the quality of the final products.
Extraction is particularly effective at moderate operating temperatures, e.g. below 50°C with CO_2, ensuring maximum stability and quality of extracted products. This is crucial for the preservation of heat-sensitive compounds, ensuring that the final product retains all its beneficial properties.

10.4 Extraction with supercritical CO_2 :

The supercritical state gives CO_2 excellent extraction power, which can be modulated at will by adjusting the operating temperature and pressure. CO_2 is a good solvent in the supercritical state, but becomes ineffective in the gaseous state. This process offers several notable advantages:

- ✓ **Chemical inertness**: CO_2 is totally inert, natural, non-toxic and inexpensive.
- ✓ **Moderate operating conditions**: Supercritical $CO_{(2)}$ operates at low temperatures and moderate pressures, with a critical temperature of 31.1°C and a pressure of 73.9 bar; this preserves heat-sensitive compounds.
- ✓ **Easy separation and economical recovery**: At the end of the extraction cycle, the solvent (CO_2) and solute (extract) are easily separated. A simple pressure release at returns the CO_2 to the gaseous state, enabling virtually total recovery of the solvent at reduced cost.
- ✓ **Reduced operating costs**: At pilot or laboratory scale, operating costs are limited thanks to continuous CO_2 recycling.

However, extraction with supercritical $CO_{(2)}$ requires a low water content in the algal biomass, as demonstrated by work carried out by the *M2P2 Laboratory* at Aix-Marseille University as part of the SHAMASH project. What's more, although this technique is highly promising, it is still costly to install on an industrial scale.

To enhance the extraction power of supercritical $CO_{(2)}$ and extend the range of extractable molecules, a co-solvent, such as ethanol, can be added. This co-solvent can optimize the extraction of specific compounds, enabling more selective and efficient extraction. Thanks to its controlled selectivity, supercritical $CO_{(2)}$ can also be used to separate compounds after an initial extraction with another solvent, such as ethanol.

In summary, supercritical $CO_{(2)}$ extraction is a versatile, high-performance method, particularly suited to extracting sensitive compounds, while minimizing solvent residues in the final extracts. However, its large-scale implementation requires specific adjustments to manage biomass moisture and installation costs.

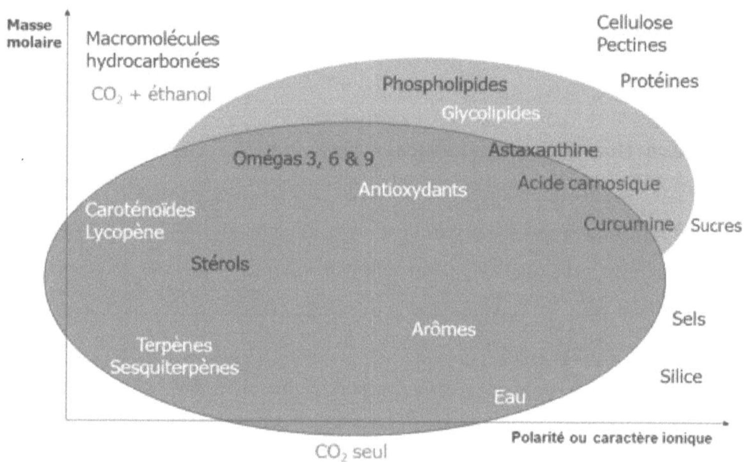

Figure 27: Molecules extractable with supercritical $CO_{(2)}$ alone and in the presence of ethanol.

11. End of process for biofuels :

11.1 Transesterification :

The transesterification reaction is a key process used to convert triglycerides extracted from algae into fatty acid methyl esters (FAME), which can be used as biodiesel. This chemical reaction involves replacing an alcohol group on an ester with another alcohol, thus transforming one ester and one alcohol into another ester and another alcohol. Transesterification can be carried out via catalytic or non-catalytic reaction systems, using various heating systems to initiate the reaction. An acid or base is often used as a catalyst to facilitate the process. This reaction produces one molecule of glycerol for every three molecules of fatty acid.

However, lipids extracted from microalgae present certain disadvantages for biodiesel production by transesterification. These lipids often contain long carbon chains and polyunsaturation, two characteristics that pose problems for biodiesel production, particularly in terms of viscosity and tar formation. By comparison, rapeseed has a fatty acid profile much more suited to biodiesel production.

11.2 Chemical transesterification :

Chemical transesterification is a relatively mature technology, often used as a reference for the conversion of vegetable oils into biodiesel. In addition to the classic approach using base-catalyzed methanol, it has been demonstrated that transesterification of algal oil can be achieved with ethanol, using sodium ethanolate as a catalyst. The products of these reactions are usually separated by adding ether and salt water to the solution, followed by thorough stirring. The biodiesel is then isolated from the ether by high-vacuum evaporation, resulting in a purified final product. In short, although chemical transesterification is a proven method for producing biodiesel from vegetable oils, adapting this technology to oils extracted from microalgae requires overcoming specific challenges linked to the composition of algal lipids.

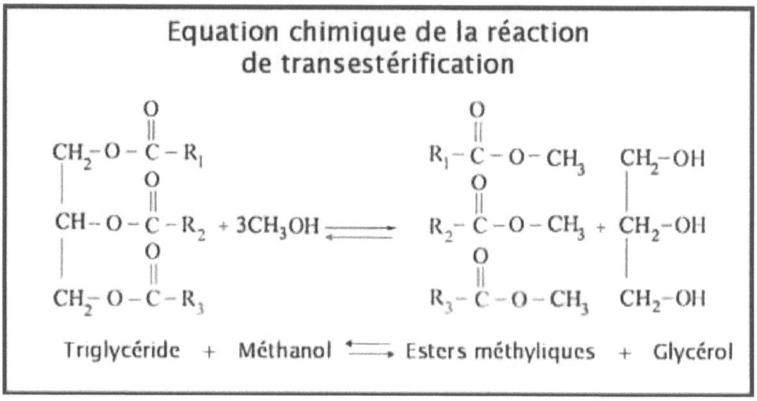

Figure 28: Classical chemical transesterification reaction.

11.3 Biochemical (enzymatic) conversion :

Chemical processes provide a high conversion of triglycerides to esters, but have drawbacks such as high energy consumption and difficulty in removing glycerol and alkaline catalyst. The use of biocatalysts (lipases) in transesterification offers a more environmentally-friendly alternative. However, although this approach is attractive, it is not yet commonly used on a large scale, mainly due to the high cost of lipases and their limited lifespan. Mastering these factors is essential to make this process economically viable.

11.4 Catalytic cracking :

Transesterification catalysts are efficient, but require purification and disposal of product streams, thus increasing costs. The development of immobilized catalysts, whether homogeneous or heterogeneous, could reduce these costs. Lewis acid catalysts, such as $AlCl_3$ or $ZnCl_2$, have been shown to be effective in converting triglycerides to methyl esters. The use of a co-solvent, such as tetrahydrofuran, can improve conversion yields, reaching up to 98%.

11.5 Hydrogenation :

Transesterification efficiently converts triglycerides into fatty acids usable as biodiesel, but these products are not suitable for bio-kerosene, which requires strict specifications. Hydrogenation, which eliminates double bonds in hydrocarbons, is an alternative for producing high-quality aviation fuels. The resulting hydrocarbons have excellent qualities for kerosene, such as well-dimensioned carbon chains, high cetane, absence of sulfur, and adjustable cold properties.

11.6 Methanization :

Methanization is an anaerobic digestion process that transforms organic matter into biogas, composed of 60-70% bio-methane, the remainder being mainly CO_2, which can be recycled for algae cultivation. Advantages of this process include the use of wet biomass, reducing the need for drying, and the ability to recycle nutrients from the digested biomass. However, challenges remain for marine species, including sulfur management and inhibition by sodium and chlorides. Although biogas is not economically viable as a primary product, methanation can be useful for processing biomass residues after extraction of high value-added products, or in processes combining organic waste treatment and microalgae cultivation.

12. Thermochemical treatments :

Thermochemical processing enables biomass to be transformed without involving living organisms, making it suitable for use in extreme conditions. Biomass is broken down into carbonaceous molecules in the gas, liquid or solid phase, which can then be refined to produce biofuels. However, this type of processing requires a high energy input, greater than that required for biogas production.

12.1 Pyrolysis :

Pyrolysis chemically decomposes biomass using heat, without oxygen or other reagents. Thermochemical processing of algae can produce a wide range of products, depending on the reaction conditions. A major advantage of pyrolysis is its speed, with reaction times ranging from seconds to minutes.

12.2 Gasification :

Gasification converts carbonaceous materials, such as biomass, into carbon monoxide and hydrogen by reacting with a controlled amount of oxygen at high temperature. The resulting mixture, called syngas, is a usable fuel. Gasification of algal biomass can produce a variety of liquid fuels, notably via Fischer-Tropsch synthesis or mixed alcohol synthesis. Fischer-Tropsch synthesis, in particular, is a mature technology, although water removal remains a challenge for algal gasification.

12.3 Hydrothermal liquefaction :

Hydrothermal liquefaction in subcritical water converts wet algal biomass into liquid fuels. This process, inspired by natural fossil fuel formation processes, exploits the high activity of water in a subcritical environment to break down biomass into smaller molecules. The main product of this process is a "bio-crude" representing around 45% by weight of the raw material, with an energy content comparable to diesel.

13. Applications and potential markets :

13.1 Introduction :

The diversity of microalgae species and their wealth of metabolites open the way to a wide range of applications, covering many industrial sectors (see figure 29 and table 4 below). Biomass can be used in its raw form or as a vector for the production of molecules of interest. Raw biomass, whether living or not, finds outlets in the human nutrition market, the animal feed market, particularly in the aquaculture sector, as well as in the energy and environmental sectors.

The compounds of interest, mainly identified in microalgae and cyanobacteria, include lipids, vitamins, polysaccharides, proteins, pigments, antioxidants and various other cellular components. These products are marketed in several sectors:

- ✓ **Health and well-being**: Use in food (food additives and colorants), nutraceuticals, cosmetics and pharmaceuticals (biological and therapeutic active ingredients).

- ✓ **Animal feed**: Used as a potential protein meal and colorant.

- ✓ **Energy**: Use in the production of biofuels and other renewable energy sources.

- ✓ **Green chemistry**: chemical intermediates and bio-sourced materials.

The environmental field is also interested in the use of cultivated algae for their depolluting properties, contributing to ecological solutions in ecosystem management and pollution reduction.

Table 4: Commercially produced microalgae, quantities, locations, applications and market value in 2004.

Micro-algae	Annual production	Producing countries	Application and products	Price
Spirulina	3,000 tonnes of dry matter	China, India, USA, Burma, Japan	Human nutrition, Animal nutrition, Cosmetics, Phycobiliproteins	36 €/kg, 11 €/mg

Chlorella	2,000 tonnes of dry matter	Taiwan, Germany, Japan	Human nutrition, Cosmetics, Aquaculture	36 €/kg, 50 €/L
Dunaliella salina	1,200 tonnes of dry matter	Australia, Israel, USA, Japan	Human nutrition, Cosmetics, B-carotene	215-2150 €/kg
Aphanizomenon flos-aquae	500 tonnes of dry matter	USA	Human nutrition	Not specified
Haematococcus pluvialis	300 tonnes of dry matter	USA, India, Israel	Aquaculture, Astaxanthin	50 €/L, 7150 €/kg
Crypthecodinium cohnii	240 tonnes of DHA oil	USA	DHA oil	43 €/kg
Shizochytrium	10 tons of DHA oil	USA	DHA oil	43 €/kg

Setting aside the challenges associated with industrial production of microalgae, the market potential for products derived from microalgae is growing significantly, in line with the global market for renewable chemicals, which was valued at 45 billion euros in 2009 and 59 billion euros in 2014. According to a study by Pulz and Gross, the global market for microalgae was estimated at between 5 and 6.5 billion US dollars (TecKnowMetrix study for Algasud - 2010).

These figures illustrate the growing interest in products derived from microalgae, as well as their commercial potential, which could continue to grow as production technologies become more efficient and cost-effective on a large scale.

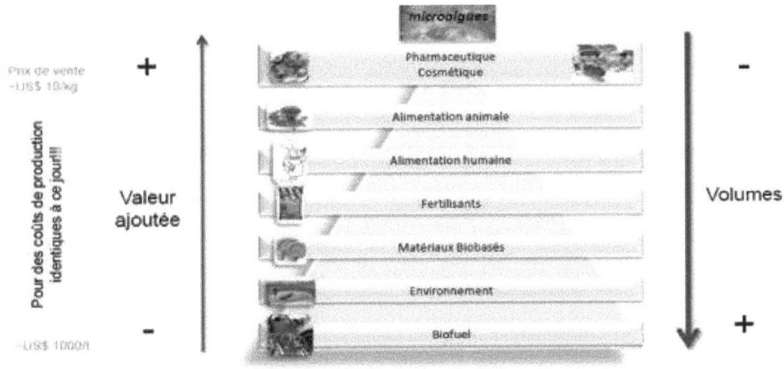

Figure 29: Microalgae, the field of possible applications.

13.2 Seaweed products for human consumption :

Seaweed is traditionally consumed in many parts of the world, particularly by indigenous peoples in Africa, South America and Mexico, who include it in small quantities in their diet to benefit from its vitamin and nutrient content. In addition to these traditional uses, ingredients derived from seaweed are increasingly used in industrial food preparation. These ingredients include polyunsaturated fatty acids, natural pigments, antioxidants and various bioactive compounds. Microalgae for human nutrition are now marketed in various forms: tablets, capsules, suspensions and soluble extracts in ampoules. They can also be incorporated into food products such as pasta, snacks, sweets, drinks, breakfast cereals, nutrition bars and cookies.

In 2007, the global market for functional foods and nutraceuticals was estimated at 138 billion euros, and was forecast to reach 175 billion euros in 2012, with a share of 1 to 2 billion euros devoted to microalgae (TecKnowMetrix study for Algasud). According to the *National Algal Biofuels Technology Roadmap*, worldwide consumption of algal biomass in the form of food supplements represented a market of around $2.5 billion in 2010, or around €1.8 billion (IEED Green Stars Project).

13.3 Staple foods :

Proteins are essential to human nutrition, and protein deficiency is one of the main causes of malnutrition. Some algae, such as the cyanobacterium **Arthrospira** (better known as *Spirulina* or spirulina), contain up to 60% protein. Spirulina, cultivated for its high protein content, is easy to grow in monoculture due to its ability to grow at high pH and form spiral-shaped multi-cellular groups, making it easy to harvest. Evidence of spirulina consumption dates back to the 16th century among the Aztecs in Mexico and the Kanembou tribe in Chad.

In 2008, a resolution entitled "Using spirulina to combat hunger and malnutrition and contribute to the achievement of sustainable development", led by five developing countries at the United Nations General Assembly, led the FAO to publish a comprehensive report on this micro-algae. The report points to many beneficial effects on human health, but also warns that certain species of algae can produce toxins, such as microcystins.

Spirulina, rich in protein (over 60% crude protein), vitamins, minerals and other biologically active substances, is easily digestible thanks to its polysaccharide cell wall, with a digestibility of 86%. A typical daily intake of spirulina is around 15 grams, representing a small proportion of an adult's daily protein intake.

There are two main spirulina extracts: phycobiliproteins, the only blue food coloring of natural origin, and a tasteless, odorless, white-yellow protein extract, which can be used for a variety of food applications. The biomass remaining after extraction can be used for other products.

Although microalgae such as chlorella are also recognized for their high protein content, they have not yet acquired significant importance as a food substitute due to regulatory constraints and organoleptic characteristics (texture, color, odor).

13.4 Health foods and pharmaceutical products :

In addition to their use as food, algae provide a wide variety of vitamins, nutraceuticals and other nutrients. Health food products currently dominate the microalgae market.

In 1996, consumption of health foods derived from microalgae in Japan was estimated at 2,400 tonnes.

Microalgae are a valuable source of pigments, polyunsaturated fatty acids (PUFAs) and other bioactive compounds with medical or nutritional properties. The global market for colorants, all applications combined, was worth 14 billion euros in 2004, and the market for carotenoids, mainly derived from microalgae, represented around 900 million US dollars in 2005.

The cosmetics industry is also a growing user of algae extracts, notably for anti-aging creams, emollients and sun protection products. Algae provide antioxidant, moisturizing and protective benefits, making them sought-after ingredients in skincare and beauty products.

13.5 Algae for animal feed :

Protein-rich algae are used in animal feed, particularly for land-based livestock (pigs, poultry, cattle) and aquaculture, as a substitute for oilseed cakes and fishmeal. According to US DEO 2010, microalgae for aquaculture and animal feed represent 540 and 230 million euros respectively.

The use of algae in animal feed offers many advantages, such as improved animal fertility and health, and better quality by-products for human consumption, such as meat and milk.

13.6 Cosmetics :

Microalgae, notably ***Arthrospira*** and ***Chlorella***, are well established in the skin care market. Their extracts, with their antioxidant, moisturizing and protective properties, are used in anti-aging creams, suncare products and hair care products. The market for natural and organic cosmetics is booming, and micro-algae offer great potential for meeting this growing demand.

13.7. Energy :

Microalgae are a promising solution in the race for renewable energies, particularly for biofuel production. They are capable of accumulating more than 50% of their dry weight in fatty acids, with yields per hectare higher than those of terrestrial oleaginous species. Biofuels derived from microalgae, such as gasoline, diesel and kerosene, offer a sustainable alternative to traditional fossil fuels, with characteristics tailored to the requirements of the transport sector, including aviation.

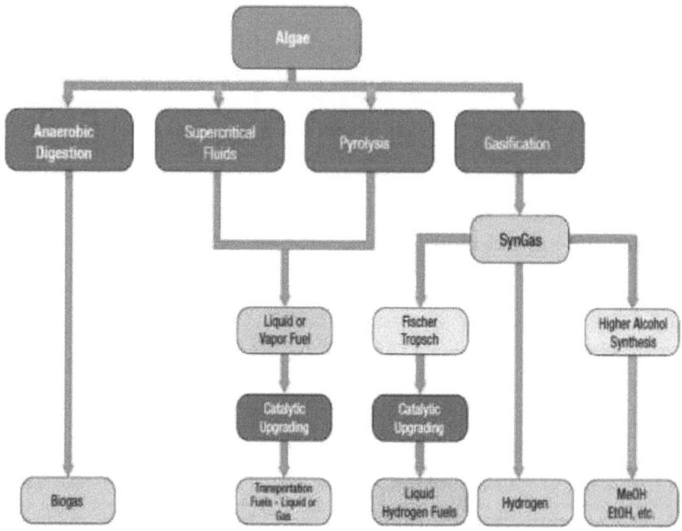

Figure 30: Schematic diagram of the potential conversion of algae in their entirety into biofuels.

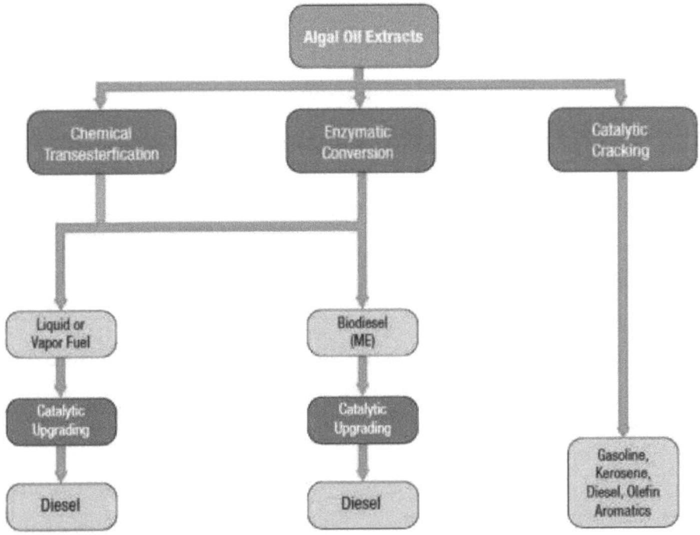

Figure 31: Diagram of different strategies for converting algae extracts into biofuels.

13.8 Biodiesel and Hydrocarbons :

Vegetable oils and animal fats are renewable resources whose chemical structure of lipid chains is similar to that of the hydrocarbons present in petroleum-derived "diesel cuts". For biodiesel production, it is crucial to use oils that are not in direct competition with food use, such as those from jatropha or algae crops.

The production of biodiesel from algal oils has attracted renewed interest over the last decade, mainly due to the remarkable productivity of certain oilseed species. Indeed, under certain deficiency conditions, algal biomass can contain over 50% lipids, whereas for plants such as rapeseed, these lipids represent only around 6% of the total plant.

However, under normal growth conditions, the concentration of triglycerides in algae is significantly lower.

To stimulate carbon storage in the form of lipid reserves, a deficiency in certain minerals, such as phosphorus, nitrogen or silicon, is necessary. Unfortunately, this deficiency leads to a reduction in the growth rate of algae, which considerably reduces overall yields.

One species of algae, *Botryococcus*, is capable of producing long-chain hydrocarbons, which are not suitable for biodiesel production but can be converted into bio-kerosene. However, *Botryococcus* has a low growth rate, which limits its productivity, and it is highly sensitive to competition from other micro-organisms, which is a further obstacle to its large-scale use.

Table 5: Comparison of some biomass sources of biodiesel.

Culture	Oil yield (L/ha)	Land area required (M ha) [a]	Percentage of existing farmland in the United States [a]
Corn	172	1540	846
Soybeans	446	594	326
Canola	1190	223	122
Jatropha	1892	140	77
Coconut	2689	99	54
palm oil	5950	45	24
Microalgae [b]	136 900	2	1,1
Microalgae [c]	58 700	4,5	2,5

[a] To meet 50% of the United States' transportation fuel needs.
[b] 70% oil (by weight) in biomass.
[c] 30% oil (by weight) in biomass.

While research is currently focused on oil-rich algae species in laboratories, it is essential to evaluate these processes on a larger scale. The aim is to go beyond the few pilot experiments carried out, mainly in the United States, to achieve production at a competitive cost. This cost must be compatible with the price of a barrel of oil, while offering a favorable energy balance and reducing greenhouse gas emissions compared with first-generation agrofuels.

Apart from the technical challenges, which are currently being resolved, the real stakes lie in mastering the economic equations and the ability to produce on a large scale. To give an order of magnitude, one hectare of microalgae cultivation can theoretically produce between 60 and 300 barrels of oil equivalent per year, compared with just 7 barrels for rapeseed.

In the oil industry, the million barrel is often used as a unit of measurement. For example, to produce one million barrels of oil equivalent per year from microalgae, a cultivation area of between 1,600 and 3,400 hectares would be required. By comparison, it would take over 140,000 hectares to achieve the same production with rapeseed.

Figure 32: The algaeus, the first algal biofuel car legally approved in the USA.

In 2009, the global biodiesel market generated sales of US$8.6 billion, corresponding to production of 12 million tonnes of oil equivalent (TOE). In Europe, this market represented 7.8 million TOE, including 2.2 million in France. Like European production, which has grown by 35% over the past 20 years, the biodiesel market continues to grow in response to environmental challenges. Europe has set each member state a target of 10% renewable energy in transport sector energy consumption by 2020. The French authorities have set themselves the ambitious target of reaching this threshold as early as 2015. Liquid biofuels, such as biodiesel, play a crucial role in meeting these targets. The profitability of biofuel production is closely linked to the price of fossil fuels, which in turn is indexed to the price of oil. Currently, crude oil is valued at €0.4/L, while rapeseed oil costs €1/L.

By 2050, the price of a barrel of oil is set to double, from €70 to €140, enabling biofuel production sectors to be gradually developed on a more sustainable economic basis.

13.9 EADS 100% algae :

In June 2010, at the Berlin Air Show (ILA), EADS achieved the feat of flying a Diamond DA42NG (new generation) every day on biofuel produced exclusively from algae. For this demonstration, 100 kilos of algae were needed to extract 22 liters of oil, enabling 21 liters of biofuel to be produced. During its growth phase, this quantity of algae would have absorbed 182 kg of CO_2, and the biofuel obtained contained eight times less fossil hydrocarbons than petroleum-derived kerosene. What's more, these flights achieved fuel savings of 5-10% without the need for specific engine adaptations. EADS estimates that by 2030, biofuels could account for up to 30% of aviation fuels.

Today, global kerosene consumption stands at 200 million tonnes, and is expected to rise to 300 or 350 million tonnes over the next twenty years. The 190 member countries of the International Civil Aviation Organization (ICAO) have set themselves the ambitious target of capping their aviation-related emissions by 2020. However, the availability of terrestrial biomass for biofuel production is largely insufficient to meet these targets. This limitation highlights the importance of developing alternatives, such as algae-based biofuels, to meet the growing needs of the aviation sector, while respecting environmental commitments.

13.9.1 Ethanol :

Ethanol is a biofuel commonly produced from feedstocks containing sugars or starch, and some algae can contain over 50% starch. The cell walls of algae are composed of polysaccharides, which can be used as feedstock in a process similar to cellulosic ethanol production. This has the added advantage that algae rarely contain lignin, and their polysaccharides are generally easier to break down than woody biomass.

Some algae, such as *Chlorella vulgaris* and *Chlamydomonas perigranulata*, can produce ethanol and other alcohols by self-fermentation of their starch. This process can be achieved by using photosynthesis to produce and store starch in the algae, or by feeding the algae sugars directly, then anaerobically fermenting these carbon sources in the dark to produce ethanol. If ethanol and other alcohols could be extracted directly from the algae culture medium, this would reduce the necessary investment and energy consumption compared with other algae-based biofuel production processes.

13.9.2 Biogas :

As far as biogas production is concerned, anaerobic digestion of microalgae is a highly promising method for producing gaseous biofuels. To date, it is probably the most technically and economically advanced method of converting microalgae into energy. This method makes it possible to harvest wet biomass that can be directly methanized by biodigestion, thereby eliminating several of the obstacles responsible for the high costs associated with algal biofuels, such as drying, extraction and fuel conversion. This approach is particularly effective for integrated wastewater treatment, where algae are grown under uncontrolled conditions using strains not optimized for lipid production.

13.9.3 Dihydrogen (H_2) :

Dihydrogen (H_2), considered to be the clean fuel of the future, presents major production challenges. Biological dihydrogen production technologies, such as photosynthetic hydrogen, photo-fermentation and obscure fermentation, offer interesting prospects.

Hydrogen production from microalgae is particularly promising, as it could capture CO_2 as biomass while producing hydrogen from sunlight, CO_2 and water. However, major technological challenges must be overcome before this method can be considered viable on a large scale, including improving $H_{(2)}$ production and dissociating this production from stress conditions for the algae.

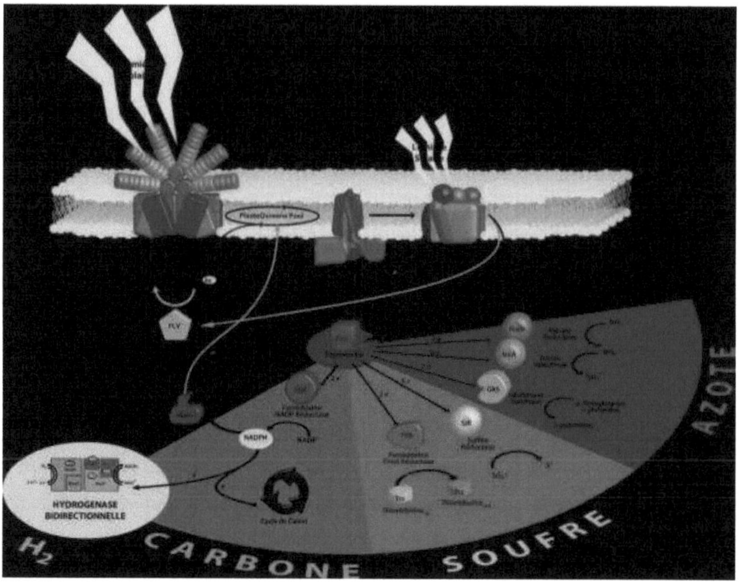

Figure 33: The different vital metabolisms of algae, dependent on photosynthesis, "competing" with photosynthetic dihydrogen production.

The future of biological hydrogen production depends on several key factors. On the one hand, scientific advances such as efficiency improvements using genetically modified microalgae or the development of advanced photobioreactors will play a crucial role. These technologies can potentially increase the capacity of microalgae to produce hydrogen while minimizing production costs. However, scientific progress alone will not be enough to ensure the viability of this sector.

Economic considerations are also essential. Hydrogen production from microalgae must be competitive with other methods of hydrogen production, such as methane reforming or electrolysis. Production, infrastructure and maintenance costs must be affordable if this technology is to be adopted on a large scale.

Social acceptability is another decisive factor. Society needs to be informed and convinced of the environmental and economic benefits of hydrogen produced from microalgae. This includes awareness-raising campaigns on the importance of renewable energies and the specific benefits of microalgae as a source of clean hydrogen.

The development of a solid hydrogen infrastructure throughout the country is essential. This includes the creation of distribution networks, refuelling stations and hydrogen storage systems. Without adequate infrastructure, even the most efficient hydrogen production technologies risk remaining under-utilized

At the same time, microalgae could be integrated into existing industrial processes, such as the production of hydrogen by reforming methane. In this context, a microalgae culture could be used to capture the CO_2 emitted by the process, thereby reducing the overall carbon footprint. In addition, the resulting algal biomass could be used to produce biogas, which could then be re-injected into the reforming process, creating a more sustainable biomass and biogas production cycle.

13.9.4 Bioelectricity :

As far as bioelectricity is concerned, although the co-combustion of algal biomass in power plants is possible, this method is generally not considered economically viable. The process of drying the biomass is energy-intensive, which reduces the overall efficiency of this recovery route. This option could nevertheless be considered in specific cases where biomass already needs to be dried for other reasons, such as the extraction of valuable co-products. In general, however, this approach is not very attractive when compared with other methods of adding value to algal biomass.

13.9.5 Environment :

The use of microalgae in the environmental field is part of a depollution strategy aimed at transforming waste into valuable resources. For example, they are used to treat wastewater rich in nitrates and phosphates, to purify gaseous effluents containing CO_2 and various oxides such as NOx and SOx, and also offer potential solutions for the bioremediation of polluted sites.

13.10 Effluent treatment :

Effluent treatment using microalgae is a practice that dates back several decades. This method, known as lagooning, is based on an extensive wastewater purification system that mimics the natural process of biodegradation in an aquatic environment. Microalgae play a key role in this method, absorbing nutrients and facilitating the degradation of pollutants, thus contributing to water depollution.

13.11 Lagoon operating principle :

The operating principle of lagooning is based on a series of successive stages designed to treat wastewater in a natural way. Before the water enters the first tank, it is pre-treated to optimize the efficiency of subsequent stages. In this first basin, bacteria play a crucial role in eliminating organic matter, transforming it into mineral salts and gases. These elements are then recovered in the second tank by micro-algae, which use these mineral salts and gases for their development. The phytoplankton thus formed is consumed in subsequent tanks by zooplankton. After a complete 80-day cycle, the treated water is ready for discharge into the Thau lagoon. The nitrogen and phosphorus present in the effluent are absorbed by micro-algae to produce energy. Agricultural effluents, such as slurry and manure, as well as those from the agri-food industry, which are rich in nitrogen and phosphorus, can thus be valorized.
Treating this effluent reduces the nitrogen content of the water, thus limiting the amount of nitrogen to be spread on farmland, in accordance with the Code of Good Agricultural Practice, which sets this limit at 170 kg per hectare.

13.12 The CO_2 market :

The $CO_{(2)}$ market is particularly relevant in the context of high-emission industries, such as thermal power plants, cement plants and refineries. These facilities are major sources of carbon dioxide (CO_2), and microalgae, by absorbing this carbon to produce energy, offer not only environmental benefits but also economic opportunities for polluting companies. In France, over 1,160 industrial sites are covered by the Kyoto

Protocol's quota system, which aims to reduce carbon emissions on a European scale. Between 2008 and 2012, French companies were allocated a total of 132 million tonnes of CO_2. According to the European Union Emissions Trading Scheme (EU ETS), the current price of a tonne of surplus carbon is 13 euros. If all French manufacturers were to sell their carbon credits, potential revenues would amount to 1,716 billion euros. The market for CO_2 capture and storage is closely linked to the technology employed, with average costs currently estimated at between 25 and 50 euros per tonne.

13.13 Bioremediation of polluted sites :

Bioremediation of polluted sites represents another crucial aspect of the use of microalgae. Recent research has shown that sites contaminated by hydrocarbons harbor consortia of cyanobacteria capable of degrading petroleum components. These cyanobacteria facilitate the degradation process by providing associated bacteria which, together with the necessary oxygen and nitrogen, effectively degrade the oil. In Europe, soil decontamination, mainly of petroleum products, has generated a market worth 30 billion euros over the last ten years.

13.14 Green chemistry :

13.14.1 Chemical industry :

The chemical industry, heavily dependent on fossil oil for the extraction of chemicals, plastics and fuels, is beginning to explore biological alternatives. Despite some similarities between the fuel and chemical industries, such as the low cost of fossil raw materials and the use of hydrocarbons, the replacement of these resources by products of biological origin remains at an early stage due to their specificity and cost, which is still too high.

With sales of 85.8 billion euros and a workforce of 182,140, the French chemical industry is the world's 5th largest producer and 3rd largest exporter. In Europe, it ranks

2nd after Germany, accounting for 15% of European chemical sales. Export sales total 44.6 billion euros. As part of the Grenelle Environment Round Table, the French chemical industry has committed to increasing the use of plant-based raw materials from the current 5% to 8% to 15% by 2017, with a target of 50% by 2050, thus supporting the development of green chemistry.
There are a number of specialized products and chemicals that can be derived from microalgae, for use in green chemistry. These include bio-flocculants, bio-polymers, synthons, biodegradable plastics, lipids, surfactants and polysaccharides. Some projects are still at the R&D stage, such as the production of bioplastics incorporating micro-algae (Cereplast, USA) or the production of paints based on molecules extracted from micro-algae (Algicoat, Netherlands).

13.14.2 Bioplastics :

Cyanobacteria, in particular spirulina, offer a promising example of biotechnological application in bioplastics production. These micro-organisms have the capacity to accumulate polyhydroxyalkanoates, polymers that can replace non-biodegradable petrochemical plastics.
The global bioplastics market was estimated at 1.1 billion euros in 2015, with demand projected at 2 million tonnes, or 5 billion euros, by 2018. In Europe, annual demand for bioplastics is expected to reach 5 million tonnes by 2020. BASF, Europe's leading market player, anticipates annual growth of 20% for bioplastics and biodegradable plastics over the next few years. This trend is confirmed by a report from the European Bioplastics association, which forecasts a 6.5-fold increase in the volume of the bioplastics market between 2007 and 2013.

13.14.3 Agro-surfactants :

Surfactants are amphiphilic molecules composed of a hydrophilic and a hydrophobic part. Because of their wetting, solubilizing, detergent and emulsifying properties, they are widely used in household and industrial detergents, as well as in cosmetics. In

2008, the global surfactants market totalled 10 million tonnes, including 2.5 million in Europe and 0.4 million in France. The lipochemistry sector in France alone generated sales of 3.7 billion euros that same year. With a view to sustainable development, and in view of the constraints imposed by the REACH directive, which requires the withdrawal of certain chemicals from the market, manufacturers are increasingly turning to agro-surfactants derived from plant-based raw materials, as a replacement for surfactants of petrochemical origin. The agro-surfactants market, in which Rhodia is a leader, is growing by an estimated 7.5% over the periods 2005-2015 and 2015-2030.

Carbon chains derived from microalgae represent a potential new source for the manufacture of bioplastics and agro-surfactants, complementing those derived from terrestrial plants. However, to promote the use of microalgae in these applications, it will be essential to adapt cultivation, harvesting and transformation processes to ensure a neutral or even negative carbon balance.

13.15 Conclusion :

At present, it is difficult to determine precisely what market share each potential sector could occupy in the application of microalgae. This uncertainty lies mainly in the expectation of stable, high-quality and economically viable biomass production, a goal that has not yet been fully achieved. To date, the most advanced applications are to be found in niche markets, characterized by annual production limited to a few kilos or tons, but with very high added value, such as cosmetics, pharmaceuticals and nutraceuticals, for both animal and human consumption. The intermediate market, which could serve as a springboard to larger-scale development, is that of animal feed, particularly in fish farming, where microalgae could replace fish meal and fish oil.

To target mass markets such as fuels and chemicals, it is imperative that the microalgae industry is organized around a central objective: reducing production costs. This also implies guaranteeing a stable supply of significant quantities of microalgae, an essential condition if these sectors are to develop and become economically viable on a large scale.

14. Regulations :

As with all other natural resources used for industrial purposes, the development of microalgae is subject to a rigorous regulatory framework.

14.1 Feeding :

Authorizations for the use of microalgae in human foodstuffs are governed by EC procedure 258/97 on "Novel Foods" (Official Journal of the European Communities, 27/01/1997). This European regulation, also known as "Novel Food", applies to foods and food ingredients composed of or isolated from micro-organisms, fungi or algae.
It is particularly relevant for seaweeds that have no proven traditional use in a member state of the European Union. In such cases, an authorization application must be submitted for the use of these algae as a food or food ingredient, in accordance with the provisions of the above-mentioned EC regulation.
It is important to note that the authorization does not only concern the species in question, but also its production method (such as outdoor raceways, natural marine environments, or specific conditions). Consequently, any change in the production method requires the submission of a new dossier to demonstrate that the product obtained is "equivalent" to that initially authorized.
Currently, only three micro-algae are authorized for human consumption in the European Union. The first is *Arthrospira platensis*, commonly known as spirulina, which has been authorized in Europe since 1981. In 2002, the diatom *Odontella aurita* received a favorable opinion from AFSSA following a request for substantial equivalence submitted by Innovalg, which enabled it to obtain authorization as a food ingredient. Finally, in 2004, France authorized the use of *Chlorella vulgaris*, an algae belonging to the chlorophyceae family.
We should also mention two extracts of marine micro-organisms (which, strictly speaking, are not micro-algae but marine fungi) authorized at European level:

- ✓ The oil extracted from *Schizochytrium sp.*, with its high DHA content (32% DHA), was granted marketing authorization under EC regulation 258/97 by the European Community in 2003.
- ✓ Oil extracted from *Ulkenia sp.* (32% DHA) was also approved for food use in 2009.
- ✓ Two pigments extracted from seaweed are also authorized as additives in the colorant category under EU regulations:
- ✓ β-carotene (E160a) from the micro-algae *Dunaliella salina* is authorized for human consumption.
- ✓ Astaxanthin (E161j), extracted from the micro-algae *Haematococcus pluvialis*, is mainly used in the feed of fish such as trout and salmon, but was already marketed as titrated extracts in dietary supplements in Europe before 1997.

14.2 In cosmetics :

At present, there are no specific regulations governing the use of algae in the cosmetics sector. The use of microalgae must therefore comply with existing regulatory constraints and official recommendations applicable to cosmetic products. This includes compliance with authorized claims, necessary certifications, REACH recommendations and cosmetics-specific nomenclatures.

In order to improve regulations in this field, several European countries have joined forces to propose good manufacturing practices through the ISO 22 716 standard. These efforts are also aimed at establishing stricter regulations for cosmetic products, scheduled to come into force in July 2013 under the REACH regulation (EC n° 1223/2009).

In addition, ISO/TC 217 and ISO/NP 16 128 have been developed to define the technical criteria and definitions of "natural" and "organic" ingredients and products in cosmetics. These standards aim to provide a more precise framework for the use of ingredients of natural origin, such as micro-algae, in the cosmetics industry.

Table 6: Examples of INCI nomenclature for microalgae-based cosmetic extracts.

INCI name	CAS NO.	Chemical name / IUPAC
Chlorella emersonii extract	223749-78-8	*Chlorella emersonii* extract is an extract of the algae *Chlorella emersonii*, Oocystaceae.
Chlorella extract	N/A	*Chlorella* ferment is an extract of the product resulting from the fermentation of *Chlorella* by yeasts.
Chlorella pyrenoidosa extract	223749-81-3	*Chlorella pyrenoidosa* extract is an extract of the *Chlorella pyrenoidosa* plant, Oocystaceae.
Chlorella vulgaris extract	223749-83-5	*Chlorella vulgaris* extract is an extract of the algae *Chlorella vulgaris*, Oocystaceae.
Porphyridium cruentum extract	223751-77-7	*Porphyridium cruentum* extract is an extract of *Porphyridium cruentum*, a seaweed.
Porphyridium zinc extract	N/A	*Porphyridium* / Zinc ferment is an extract of the product of *Porphyridium* fermentation in the presence of zinc ions.
Spirulina maxima extract	N/A	*Spirulina maxima* extract is an extract of the fronds *of Spirulina*, *Spirulina maxima*, Cyanophyceae.
Spirulina platensis extract	223751-80-2	*Spirulina platensis* extract is an extract of the *Spirulina platensis* algae, Cyanophyceae.

A general description of the cosmetic extract is all that's needed to introduce extracts from micro-algae into beauty products, with no obligation to specify the species of algae used to produce the extract. This practice is facilitated by the INCI nomenclature (International Nomenclature of Cosmetic Ingredients), which was initiated in 1973 by the C.T.F.A. (Cosmetic, Toiletry and Fragrance Association) and made mandatory in

Europe in 1998. This nomenclature provides consumers with information on the composition of ingredients, listed in descending order of concentration, under internationally recognized names.

It's also worth mentioning COSMED, an association dedicated to helping manufacturers meet regulatory requirements, which plays a crucial role in ensuring that cosmetics comply with current standards.

14.3 More generally :

The joint AFSSA and AFSSET report published in 2006 on "Assessing the health risks associated with the presence of cyanobacteria in water" also addressed the risks associated with the consumption of certain food supplements containing cyanobacteria.

In particular, the report warns against the use of products based on Aphanizomenon flos-aquae var. flos-aquae, which are sometimes marketed in Europe as spirulina equivalents. Some batches of these products have been found to contain concentrations in excess of 2 µg of microcystins per gram, representing a health risk.

Bodies such as the World Health Organization (WHO) are recommending stricter controls for dietary supplements based on micro-algae and cyanobacteria, to prevent confusion between species and avoid contamination that could lead to consumer exposure to cyanotoxins or phycotoxins. Studies carried out by health organizations have identified the resources available to carry out these controls, not only on drinking and bathing water, but also, indirectly, on food supplements.

Eventually, these health guidelines could also be extended to controls on microalgae and cyanobacteria cultivation in semi-controlled production systems, even for non-food or non-parapharmaceutical applications. The aim would be to prevent the spread of toxin-producing strains in the working environment, thereby guaranteeing the safety of microalgae cultivation operators, as well as preventing any pollution of the natural environment.

CHAPTER 3: MACRO-ALGAE

1. Introduction :

Macro-algae have been consumed for millennia in Asia, notably in Japan where the first traces of their use in food [date] back to the 4th century, and have also been used in traditional Chinese medicine. In Europe and America, certain coastal populations used them as early as the Middle Ages. Today, macro-algae are mainly used as human food worldwide. In Europe, however, they are considered more as a raw material for industry. In the [16th] century, they were used in glassmaking, and an ordinance by Colbert governing the cutting of kelp appeared in the following century. From the 19th [century] onwards, seaweed found its place in the pharmaceutical industry, notably with the extraction of iodine, used as an antiseptic. Then, in the early 20th [century], they were integrated into the colloid industry with the extraction of alginates and carrageenans. In France, seaweed consumption began in earnest in the late 1970s, driven by vegetarian and macrobiotic consumers and the arrival of migrant Asian populations. At the time, seaweed was mainly imported from Asia, but entrepreneurs in Brittany were able to take advantage of the region's rich seaweed resources to offer processed products made from local seaweed. As early as the 1980s, companies began harvesting and processing seaweed for a variety of applications. Worldwide production of macro-algae currently stands at 14.3 million tonnes gross, mainly in the form of whole algae pre-dried on site to 70% dry matter. This macro-algal biomass market, worth an estimated 8 billion US dollars, has experienced significant growth, particularly over the last decade, due in large part to China's entry into the competition with the development of vast aquaculture facilities. Today, 90% of the world's production comes from aquaculture, 70% of which is carried out in China, representing 10.1 million tonnes worth 5.2 billion US dollars. The rest of the world's production is also concentrated in Asia, with the Philippines (1.5 million tonnes), Indonesia (0.91 million tonnes), the Republic of Korea (0.77 million tonnes) and Japan (0.49 million tonnes). Japan is the world's second-largest producer of aquatic plants in terms of value, thanks to its production of Nori (*Porphyra sp.*), a seaweed with a high market value, estimated at 1.1 billion US dollars.

Japanese Konbu kelp (*Laminaria japonica*), with 4.9 million tons, is the most widely cultivated seaweed, followed by Wakame (*Undaria pinnatifida*) with 2.4 million tons and Nori (Porphyra sp.) with 1.5 million tons.

France ranks 9th in the world with annual production of around 90,000 tonnes, making it Europe's second largest producer, behind Norway with 150,000 tonnes. Seaweed production and processing in France is almost exclusively concentrated in Brittany, particularly in Finistère and Côtes d'Armor, home to Europe's largest seaweed beds. On average, 70,000 tonnes of seaweed are harvested each year, mostly wild seaweed, mainly brown seaweed for the alginate market. Macroalgae are a global resource, with significant import and export flows. Although China is the world's largest producer, it is also a major importer. Europe is gradually tending to import its algal resources from other parts of the world, while countries in South America and Southeast Asia are positioning themselves primarily as biomass producers and major exporters.

Macro-algae are used in the production of foodstuffs, animal feed, chemicals, cosmetics and pharmaceuticals. While most production is concentrated in Asian countries, the USA, Canada and several European countries, including France, Ireland, the UK (Scotland, Northern Ireland, Wales), Germany, Norway, the Netherlands and Denmark, are looking to establish large-scale algae cultivation.

Table 7: Global macro-algae production.

Country	Production (million tonnes)	Value (billions of US dollars)	Comments
China	10,1	5,2	70% of world production
Philippines	1,5	-	Concentration in Asia
Indonesia	0,91	-	Concentration in Asia
Republic of Korea	0,77	-	Concentration in Asia
Japan	0,49	1,1	World's second-largest producer by value, thanks to Nori

Norway	0,15	-	Europe's leading producer
France	0,09	-	Europe's second-largest producer

2. Characteristics of macro-algae :

2.1 Diversity of families and species :

Macroalgae, while having no formal taxonomic value, generally refers to all multicellular photosynthetic aquatic organisms with the exception of terrestrial plants, known as Embryophytes. The term encompasses a wide range of organisms, distinct from micro-algae, and it's important to note that photosynthetic prokaryotes such as cyanobacteria, often referred to as blue-green algae, are not considered algae in the strict sense. Macroalgae are found in several independent evolutionary lineages, including at the base of the land plant lineage, which descended from a green alga. According to the nuclear phylogeny of eukaryotes, green and red algae share a common ancestor, while brown algae result from a secondary endosymbiosis involving a red alga, which gave rise to the plastid of brown algae. The three main macro-algae lineages, the result of a complex evolutionary history, are the Chlorophytes (green algae), the Rhodoplantae (red algae) and the Phaeophyceae (brown algae).

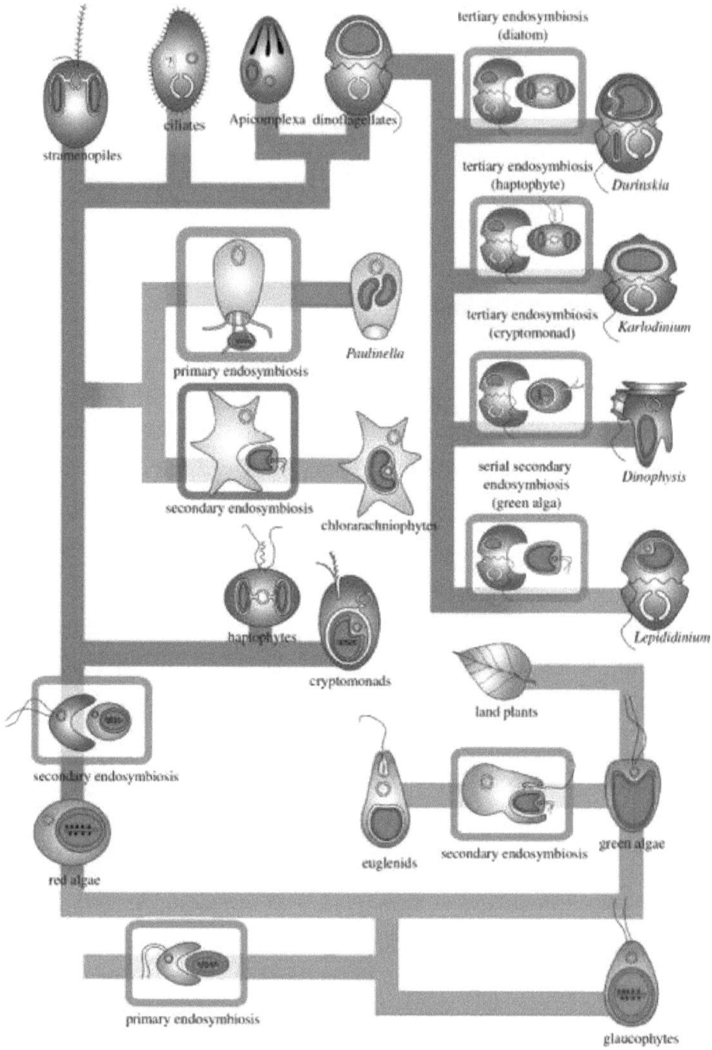

Figure 34: Diagram of the evolution of photosynthetic eukaryotic lineages.

There are currently around 1,800 species of brown macro-algae, almost 6,000 species of red macro-algae and around 1,200 species of green macro-algae. These macro-algae represent a vast group of organisms distinct from terrestrial plants, the Embryophytes, and possess a wide diversity of metabolisms, some of which are rare or even absent in animals, plants, fungi or microorganisms. This makes them valuable biological resources that remain relatively unexplored.

2.2 Biology, lifestyle and reproduction :

Macro-algae play a crucial role in coastal ecosystems, where they contribute to high genetic diversity and can colonize a variety of habitats. They sometimes form dense, extensive blankets, such as kelp "forests", which are particularly rich and productive ecosystems. These organisms are also characterized by their complex and varied reproductive cycles, which can be difficult to control. These cycles involve a succession of different stages, including haploid gametophytes and diploid sporophytes. Some macro-algae species have morphologically very distinct haploid and diploid phases, with macroscopic phases up to several meters long and free-living microscopic phases consisting of just a few cells. To add to the complexity, these reproductive cycles often include loops of parthenogenesis, and may also include modes of asexual reproduction through the development of mitotic or meiotic spores.

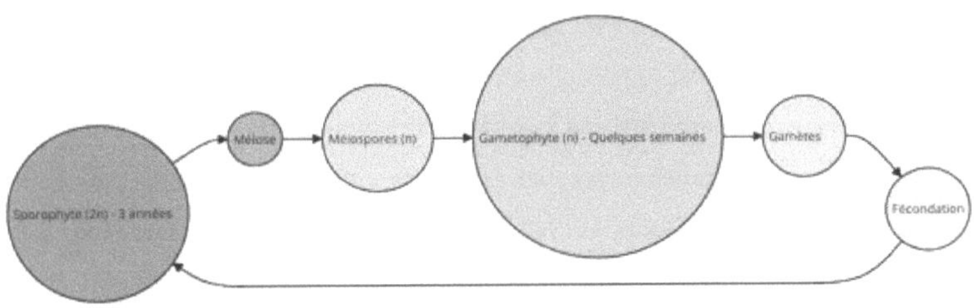

Figure 35: The reproductive cycle of the brown alga *Laminaria digitata.*

The presence of a substrate, whether natural or artificial, is crucial for the growth of algal biomass. In their natural environment, macro-algae grow mainly on rocky substrates, where they form stable, multi-layered, perennial vegetation, capturing virtually all available photons. Rocky coast algae anchor themselves firmly to rocks or pebbles with sturdy spikes, and can grow at depths of 50 to 75 meters. However, their abundance rapidly diminishes beyond 30 meters, as the radiation required for photosynthesis is absorbed by seawater. Among these algae are giant species such as kelps, some of which can reach lengths of 50 meters, depending on species and environmental conditions. As with all chlorophyllous plants, photosynthesis enables algae to fix carbon, whether atmospheric or dissolved in water, using light energy to produce biomass. Algae use a variety of chlorophyll pigments to capture photons of different wavelengths, optimizing their ability to photosynthesize in different light conditions.

2.3 Biochemical composition :

The biochemical composition of macro-algae varies considerably according to species, season and growth or stress conditions. The majority of bioactive compounds identified come from brown and red algae, while green algae remain a relatively unexplored area. The interest in macroalgae lies largely in the presence of three main categories of components: fibers, minerals and proteins, as well as metabolites with antioxidant and anti-free radical properties, such as carotenoids, polyphenols, vitamins or polyunsaturated fatty acids.

A particular feature of macro-algae, linked to their marine origin, is their high mineral and trace element content. They absorb exceptional quantities of minerals from seawater, representing up to 40% of their dry mass, particularly in the case of brown algae. These include macroelements such as sodium, calcium, magnesium, potassium, chlorine, sulfur and phosphorus, as well as trace elements such as iodine, iron, zinc, copper, selenium and molybdenum, among others.

In particular, the iodine content of macro-algae is exceptional, with certain species such as kelp (brown algae) and gracilaria (red algae) containing up to 10,000 mg/kg.

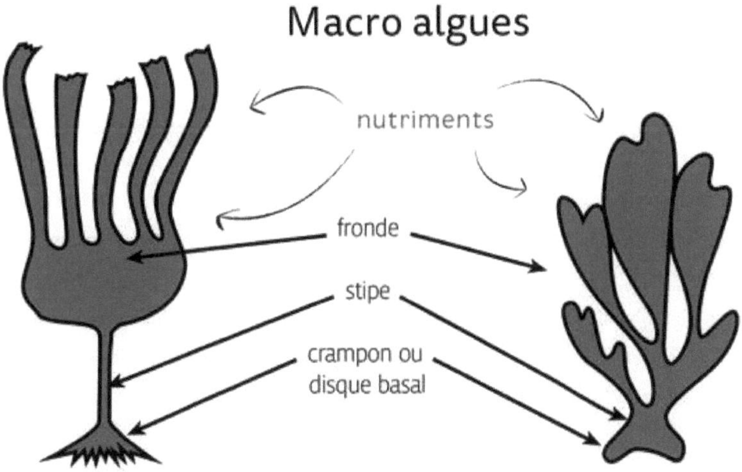

Figure 35: Schematic representation of a brown fucal macro-alga.

Protein content varies from species to species. Brown algae have a relatively low protein content (5-11% of dry matter), while some red algae can reach 30-40% of dry matter in protein. Green algae, which are not yet widely exploited, also have a significant protein content, reaching 20% of dry matter, and up to 40% in nitrogen-rich environments. Among seaweed proteins, phycobiliproteins are particularly important as pigments in the collection of light energy, and have antioxidant properties.

The lipid content of adult macro-algae is generally low, between 1 and 5% of dry matter. However, when macro-algae are maintained as gametophytes, their lipid content can reach several tens of percent, comparable to that of phytoplankton. Qualitatively speaking, algal lipids contain a high proportion of essential fatty acids, particularly polyunsaturated fatty acids in red algae. For example, the omega-3 fatty acid EPA accounts for 50% of polyunsaturated fatty acids in *Porphyra sp.* and *Palmaria palmata*.

What's more, macroalgae contain sterols in quantities similar to those found in terrestrial plants, as well as unique carotenoids such as fucoxanthin, which has highly sought-after antioxidant properties.

The vitamin composition of macro-algae is also interesting, although subject to seasonal variations and influenced by the algae treatment process. They are rich in provitamin A in red algae, vitamin C in brown and green algae, and vitamin E in brown algae. Unlike land plants, algae contain significant quantities of vitamin B12.

Most marine macro-algae also contain polyphenols, known as phlorotannins, which form a highly heterogeneous group of molecules, offering a wide range of potential biological activities. The highest levels are found in brown algae, where they represent between 5 and 20% of dry weight. Carotenoids, present on average at 0.1% of dry weight, are also powerful antioxidants, with particularly high levels in brown algae. Finally, macro-algae are distinguished by their richness in specific polysaccharides, such as phycocolloids, which account for between 18 and 45% of the dry weight of brown algae. These polysaccharides, non-digestible and absent from terrestrial plants, include alginates, carrageenans, agars, as well as other polymers such as xylans, fucans, laminarans and ulvans. They play a crucial role in cell structure and storage, and hydrocolloids derived from macro-algae are today the main industrial products derived from these organisms.

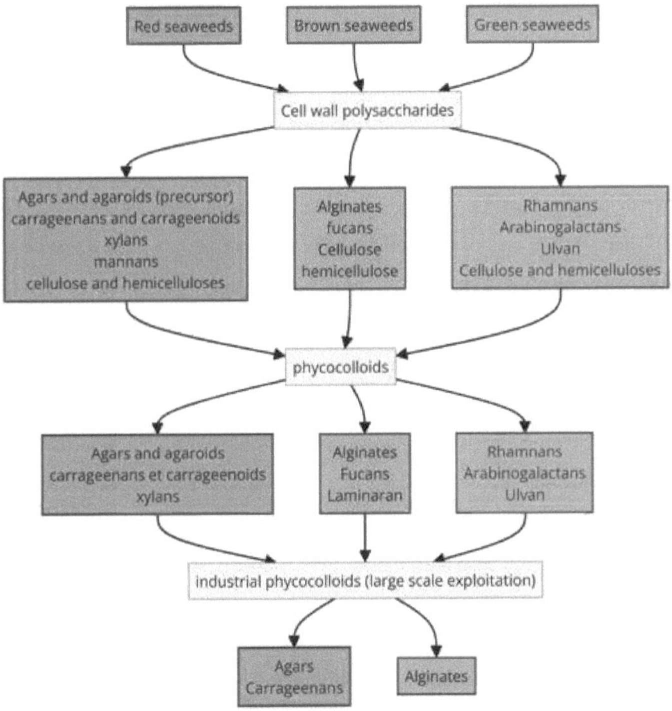

Figure 37: Main colloids present in each species family.

Industrial colloids derived from macro-algae play an essential role in various sectors. Here is an overview of the main ones:

Alginates: Alginates are polysaccharides extracted mainly from brown algae, notably kelp, but also from certain wrack species such as *Himanthalia elongata* and *Ascophyllum nodosum*. Discovered by Stanford in 1881 as part of industrial developments linked to the production of iodine from brown seaweed, industrial production of alginates began in earnest in the UK and the USA in the 1930s.

These polymers, which make up an important part of the cell wall of a wide variety of brown algae species, including *Laminaria*, *Ascophyllum* and *Macrocystis*, play a major

structural role. They are involved in the maintenance of mechanical strength, cell and tissue adhesion, morphogenesis, development, hydration and protection of algae.
Alginates are composed of two hexuronic acids: ß-D-mannuronic acid and α-L-guluronic acid. Their chemical composition varies according to seaweed species, plant parts and seasons. However, thanks to a rigorous selection of raw materials, it is possible to produce alginates with consistent characteristics. These polymers are distinguished by their lack of sulfation and low polarity.
Alginates have a wide range of technical properties, making them versatile as texturizers, thickeners, gelling agents, stabilizers, cryoprotectants (especially for frozen foods) and edible films. Their gelling properties depend on the mode of cross-linking, as well as the content and sequence of hexuronic acid monomers. For example, polyguluronic homopolymer blocks are particularly suited to gelling in the presence of calcium, producing calcium alginate gels that are hard and thermostable. Used as food additives (E400 to E405), alginates enable the reconstruction of foods, particularly fruit. Alginic acid gels, on the other hand, are soft, elastic and fusible, and can replace gelatin.
In water, alkaline alginates form viscous colloidal solutions, while calcium alginate, like alginic acid itself, is insoluble. Sodium alginate is highly soluble in water, while calcium alginate is only soluble in a basic medium. Alginates can also be converted by the chemical industry into hydroxypropyl alginates (or propylene glycol alginates, PGA), which serve as emulsifiers and stabilizers for oil-in-water emulsions, notably in ready-to-use sauces and dressings.

Carrageenans: Carrageenans are polysaccharides extracted from red seaweed, in particular Chondrus crispus, a seaweed also known as "Irish moss", native to County Carragheen in Ireland. Historically, this seaweed was used to make ointments and custards.
During the colonization of North America in the ^{18th} , the Irish discovered that this seaweed also grew on the coast of Massachusetts, which aroused the interest of industrialists due to its abundance. In 1871, the pure polysaccharide responsible for the virtues of Irish Moss was extracted in the United States and named carrageenan, in reference to its county of origin.

The first scientific work on carrageenans dates back to 1928, with a patent detailing their extraction method. After the Second World War, the food industry began to use carrageenans as stabilizers, thickeners and gelling agents, but it was only in the 1960s, following Japan's retreat from the agar market, that the carrageenan industry really took off.

Carrageenans are extracted from the cell walls of red algae using hot water. They are linear-chain polymers whose basic monomer is galactose, more or less sulfated. Smith and Cook discovered that, when an extract of *Chondrus crispus* is mixed with potassium chloride (KCl), there are at least two types of carrageenan: lambda-carrageenan, which is soluble in the presence of KCl and never forms a gel, and kappa-carrageenan, which is insoluble when cold but soluble when hot, forming a soft gel on cooling. A third type, iota-carrageenan, acts like kappa-carrageenan but produces more elastic gels. These different types of carrageenan are distinguished mainly by their degree of sulfation.

Each species of carrageenan algae has a carrageenan content that varies between 15 and 60% of the algae's mass over the course of the year, as well as by the nature of the kappa, lambda or iota fractions it contains. For example, *Kappaphycus alvarezii*, commercially known as "cottonii", provides a kappa-type extract, while *Eucheuma denticulatum*, or "*spinosum*", produces an iota-type extract. Some algae, such as *Chondrus crispus*, release lambda, kappa and iota fractions simultaneously.

Carrageenan reactivity is linked to the abundance of "OSO3-" hydrophilic groups (20 to 25%) and CH_2-O hydrophobic bridges in anhydrogalactose. The more sulfate groups a molecule contains, the more soluble it is; conversely, hydrophobic bridges reduce solubility and increase rigidity. Although carrageenans are soluble in water, they are not soluble in apolar and most polar organic solvents, except in the presence of cationic surfactants.

Lambda-carrageenan, diluted in an aqueous solution, increases viscosity without forming a gel, while iota and kappa-carrageenan act as gelling agents, iota producing soft, elastic gels, and kappa hard gels. The latter also have thickening properties when dissolved in hot water.

The gelling power of carrageenans is strongly influenced by the presence of electrolytes, essential for gel formation. Potassium salts (K^+) are the most effective in enhancing gel density, also increasing melting and gel-forming temperatures. Calcium ions (Ca_2^+) have a similar influence, particularly on iota-carrageenan, with which they form highly elastic and coherent gels. There is also a specific interaction between carrageenans and milk kappa-casein, which explains their traditional and industrial use in the dairy industry.

Carrageenans are widely used as thickeners, gelling agents and stabilizers for emulsions and suspensions in the food, textile, cosmetics (toothpastes, shampoos, lipsticks) and pharmaceutical industries. Their ability to form gels at high temperatures, up to 60°C, makes them particularly interesting compared with traditional animal gelatins.

Agars: Agars, discovered in Asia several centuries ago, take their name from the Indonesian-Malay word "agar", meaning "seaweed". Agar-agar, a natural extract from the cell wall of certain species of red algae (rhodophyceae), is composed of two main polysaccharides: agarose and agaropectin. Agarose is a neutral homopolymer made up of agarobioses, a disaccharide unit composed of D-galactose and 3,6-anhydro-L-galactopyranose. Agaropectin, on the other hand, contains both pure agarobiose units and units substituted by sulfate, methyl or pyruvate groups, among others. Agar quality is directly linked to agarose content: the higher the content, the better the gels obtained. Agarose gels are distinguished by their unique properties, offering high strength without the need for transverse binders. They are thermoreversible and thermostable, with a gelation temperature of between 32 and 43°C and a melting temperature of around 70°C. These gels are also stable to pH variations, resistant to micro-organisms and cations, and have a high water retention capacity, as well as the ability to trap hydrophobic compounds.

Like carrageenans, agars are extracted with hot water. The *Gracilaria* and *Gelidium* genera provide the majority of raw materials for agar production. These substances are mainly used as stabilizers in emulsions and suspensions, and as gelling agents. Around 90% of the agar produced is destined for food applications, while the remaining, more refined 10% is used in various biotechnological applications.

Ulvanes: Ulvanes are water-soluble, sulfated and carboxylated anionic polysaccharides extracted from the cell walls of green algae belonging to the Ulvales family, notably *Ulva* spp. and, by extension, *Enteromorpha* spp. Their extraction is generally carried out using water containing a cation chelator, and their content can vary from 8 to 29% of the dry weight of the algae. The main sugars making up ulvans include 3-sulfated L-rhamnose, D-glucuronic acid and L-iduronic acid, with minority contributions from D-galactose, D-glucose and D-xylose. Ulvans are notable as the only known examples of L-iduronic acid in the plant kingdom, a uronic acid also found in glycosaminoglycans (GAGs) such as heparin, dermatan sulfate and heparan sulfate, expressed on the surface of mammalian cells.

To date, ulvanes are not commercially available. However, they have recently been proposed for a variety of potential applications, including as a source of rare sugar precursors for the synthesis of fine chemicals, as a source of oligosaccharides for use in phytopharmaceutical or cosmetic products, and as texturizing agents for the formulation of thermoreversible gels, enabling precise control of textures.

Fucoidanes: Fucoidanes are sulfated polysaccharides present in the cell walls of brown algae. They play a crucial role in regulating hydration and osmotic stress within algal cells. Mainly composed of a repetition of sulfated fucose units, fucoidans may also contain other neutral sugars and uronic acid, sometimes acetylated. Their content varies from 3% to 30% of the alga's dry weight. In addition, fucoidans represent a potential co-product of alginate extraction, offering interesting prospects for their valorization in various industrial applications.

3. Biomass production and harvesting :

Over the last decade, the "marine macro-algae" sector has expanded significantly worldwide. This growth is largely due to the ready availability of a large biomass, making seaweed increasingly attractive for a variety of applications, including food, cosmetics, agriculture and even energy. Today, 75% of the world's algae production is destined for human consumption, mainly in the form of "vegetable algae". In France, however, the majority of harvested or imported seaweed is processed in plants for the manufacture of colloids (alginates, carrageenans, agarose) used as gelling agents and texturizers in the food industry and other specialized sectors.

In 2008, world production of fresh seaweed totalled 15.74 million tonnes. In Europe, the macro-algae industry relies mainly on harvesting algae from natural populations, with seaweed cultivation accounting for just 0.3% of volumes used. France, thanks to the Brittany region, is one of Europe's leading seaweed producers. Seaweed harvesting in Brittany is at the heart of a specific economic sector, which includes almost all French harvests and processing companies. In 2010, according to data from France AgriMer, France imported around 18,000 tonnes of seaweed (including fresh seaweed, carrageenan, agar and alginic acid) and exported 7,500 tonnes, while national production is close to 60,000 tonnes per year.

With regard to seaweed processing, a study commissioned from the CCI de Brest by the Chambre Syndicale des Algues et des Végétaux Marins (CSVAM) revealed that 67 establishments in Brittany manufacture or market products based on seaweed and marine plants, representing 1,635 jobs (excluding the agri-food sector) and generating total sales estimated at 424 million euros in 2007.

There are marked differences between Asian and Western production. In Asia, aquaculture, and more specifically seaweed farming, has grown rapidly in importance. It is estimated that the Japanese were already cultivating algae in the 5th and 6th centuries (e.g. *Laminaria japonica*), as evidenced by numerous engravings. Seaweed of the *Porphyra* genus, used in maki and sushi, began to be cultivated as early as the 18th century. Today, almost all the seaweed produced in Asia comes from cultivation, often at sea, on specialized farms. *Laminaria japonica* is the most important, with 4.2 million tons grown mainly in China.

Around 200 species of algae are used worldwide, and around ten of them are intensively cultivated, such as the brown algae *Saccharina* (*Laminaria*) *japonica* and *Undaria pinnatifida*, the red algae *Porphyra*, *Eucheuma*, *Kappaphycus* and *Gracilaria*, and the green algae *Monostroma* and *Ulva*.

Although cultivation technologies have evolved little in Asia in recent decades, the growing global demand for biomass has led to renewed investment in the search for innovations in algae cultivation to meet new business models.

3.1 Harvesting algae from natural populations and strandings :

In France, seaweed harvesting is concentrated mainly in Brittany, either by boat (on-board fishing) or on foot (shore harvesting). The seaweed collected in this way is then processed or marketed for a variety of applications, notably in the food, cosmetics and agricultural sectors.

3.2 On-board fishing for kelp :

Laminaria, particularly *Laminaria digitata*, is harvested by specialized boats in Brittany. This species is the main one used in France to manufacture alginates, and is harvested by a fleet of around thirty vessels equipped with hydraulic arms manipulating a hook called a scoubidou. Between 40,000 and 60,000 tonnes of kelp are harvested every year, generating sales of between 1.7 and 2.7 million euros. The introduction of the scoubidou in 1971, developed by a seaweed grower in collaboration with the Institut des Pêches (now IFREMER), revolutionized kelp harvesting. The hook, once submerged, rotates to roll up and pluck the seaweed from the seabed. It is then directed over the boat's deck, where a simple reversal of the direction of rotation drops the kelp onto the deck. This seaweed, located at a maximum depth of five meters below low tide, withstands natural fluctuations well, and is exploited steadily in terms of tonnage landed.

At present, some 35 licenses enable seaweed harvesters to harvest this highly regulated seasonal activity, which takes place from May to October.

Harvesting of *Laminaria hyperborea* by seaweed vessels in Brittany is more recent, having begun in the early 1990s with the first mechanized harvesting projects. Prior to this, only beached seaweed was collected. In 1994, IFREMER supervised the first mechanized harvesting trials using a technique imported from Norway, where this species is harvested from natural populations to the tune of over 100,000 tonnes per year. Today, thanks to a "Norwegian comb" modified for the Breton seabed, almost 12,000 tonnes of *Laminaria hyperborea* are harvested each year, although this activity is highly dependent on weather conditions.

Laminaria production in Brittany is subject to variations in quality, and the quantities available in the natural environment are limited. Most of this production takes place in environmentally protected areas, such as marine parks and Natura 2000 zones, and is subject to sustainable management plans to preserve the resource, biodiversity and associated ecosystem services.

Seaweed farming also raises problems due to the increasing number of conflicts of use on maritime concessions, making the bivalve/seaweed co-culture, described below, particularly relevant. In conclusion, to ensure the development of the industry with an increase in the quantities of seaweed harvested in France, it is essential to implement rigorously managed aquaculture, with regular monitoring to assess the evolution of seaweed populations and the impact on the marine environment.

Figure 38: On-board fishing for kelp with the scoubidou.

3.3 Harvesting shore algae :

The harvesting of shore algae, in particular wrack (or black wrack) and various edible seaweeds, is carried out on foot on the foreshore by professional and occasional harvesters. In 2010, it was estimated that around 50 professional harvesters were working under ENIM (Etablissement National des Invalides de la Marine) or MSA (Mutuelle Sociale Agricole) status, while 300 occasional harvesters were using TESA (Titre Emploi Simplifié Agricole) status. Ten years ago, the number of harvesters was much higher, reaching up to 1,000

In Brittany, black seaweed is still harvested by hand, using a sickle, at ebb tide. The goémonier grabs a clump of seaweed by hand and cuts it off at the base with his sickle. To preserve stock replenishment capacity, the cutting height is regulated at 20 centimeters from the base of the thallus for *Ascophyllum nodosum*, one of the main seaweeds harvested to supply industries producing seaweed extracts for agriculture or seaweed meal.

Although this type of collection has been mechanized over time in Norway, Iceland and on the east coast of Canada, thanks to the use of "faucardeur" boats adapted to the seabed, mechanization has not been able to develop on the French or Irish coasts due to the configuration of the coastline.

Hand-collected miscellaneous edible seaweed includes 12 species authorized for consumption by the French Ministry of Health for companies processing seaweed for the food industry. In some regions, such as Normandy and Finistère-Nord, these same species are also harvested for use as fodder in recently developed abalone farms. However, when it comes to assessing the impact of these new species on ecosystems, effective tools and real monitoring of harvesting are still lacking.

Figure 39: Harvesting seaweed with a sickle.

3.4 Stranding pick-ups :

Collecting natural strandings is an alternative method of collecting macro-algal biomass that can be seen as a sustainable way of managing the exploitation of wild algal populations. Historically practised on Brittany's coasts, "wrecked seaweed" was collected for the glass industry in the 17th and 18th centuries, then for iodine production in the 19th and early 20th centuries. Today, this collection method is still used on an occasional basis to meet the needs of coastal agriculture or for specific requirements, such as the harvesting of *Laminaria hyperborea* stipes for the alginate industry.

This harvesting method remains the main means of exploiting natural populations of *Macrocystis pyrifera* and *Lessonia* spp., representing 48% for *L. nigrescens* and 11% for *L. trabeculata* in Chile. On the northern Chilean and southern Peruvian coasts, quantities exceeding 100,000 tonnes of dry seaweed are collected annually. These coasts, subject to the Humboldt current which cools the surface water, are ideal for these massive strandings. After harvesting, the seaweed is dried on the beach, then either sent to crushing plants or exported directly. However, this method is restrictive and dependent on the random nature of the strandings, and is mainly practiced by coastal fishing communities or by modest populations who mobilize as soon as the strandings are exploitable to preserve the quality of the resource.

In the Basque country, particularly on the beaches of Hendaye, red algae are also harvested from strandings using bulldozers or boats equipped with nets. This biomass, composed mainly of *Gelidium sesquipedale*, provides a high-quality raw material for the extraction of agarose, one of the most expensive phycocolloids on the market, in a Basque factory.

In recent decades, the effects of coastal eutrophication, caused by massive nitrogen leakage from land-based agriculture, have become apparent. Nutrient-enriched areas of the coastline encourage the disproportionate proliferation of certain species of algae. Collecting strandings and floating algae helps reduce the negative impact of this proliferation on ecosystems. For example, between 50,000 and 90,000 tonnes of *Ulva* green algae strand on Brittany's beaches every year, and these quantities represent only the visible part of the phenomenon.

The seaweed stock at sea is estimated to be 5 to 6 times greater than the strandings, with a very rapid regeneration capacity, in just a few days under favorable conditions.

At present, little use is made of this biomass. Collected strandings are either spread directly on agricultural land, or transformed into an organic amendment (co-composting) for deferred spreading.

However, one of the main challenges of this system remains logistics and collection tools, which need to be improved to cope with massive blooms. It would be necessary to set up an organized chain for the collection, stabilization and use of these algae. However, it should be noted that the variability of strandings, linked to environmental disturbance, makes it difficult to exploit these resources on a long-term basis. It therefore seems more appropriate to adopt an opportunistic approach depending on the quantities available, rather than building a real industry on this model.

Figure 40: Collecting stranded green algae

4. Seaweed cultivation worldwide :

In 2008, global production of cultivated seaweed reached over 14 million tonnes of fresh seaweed, while only one million tonnes of wild seaweed was harvested. Most of this production is concentrated in Asian countries such as China, the Philippines, North and South Korea, Japan and Indonesia, as well as in East Africa, notably Zanzibar. The USA, Canada, and several European countries, including France, the UK (Scotland, Wales, Northern Ireland), Ireland, Germany, Norway, Denmark and the Netherlands, are also working to establish large-scale algae cultivation.

Macroalgae can be cultivated using a variety of methods, depending on the species targeted. Certain algae with a relatively simple reproductive cycle, particularly those associated with green tides, lend themselves well to cultivation in ponds or lagoons without the need for attachment infrastructures. These proliferating algae reproduce mainly vegetatively, floating freely in the water column, by fragmenting an individual. Some agarophytes, such as *Gracilaria* sp. in Chile and Asia, are also cultivated on a large scale by cuttings, *as* is *Asparagopsis armata*, a seaweed grown for cosmetics on the island of Ouessant.

The exploitation of large noble species, such as certain brown or red algae, requires prior passage through hatchery culture stages, where the algae are maintained in a microscopic state to prepare "seeds". These seeds grow attached to strings until they reach the seedling stage, when they are acclimatized to the open marine environment, where they develop into adult individuals. These adult algae continue to grow on various support devices, such as ropes or nets. This process requires in-depth knowledge and precise control of the different growth stages and reproductive cycles of algae.

A vegetative culture method is also possible in the hatchery, followed by various culture techniques, such as line, bag or pouch culture. However, the most commonly used method is string culture, which will be described in detail in the following section.

5. Rope culture stages :

First of all, seaweed is harvested by cutting in its natural environment. For kelp, the scoubidou technique on seaweed boats is often used. Once the algae have been brought back to the hatcheries, they are subjected to water, heat and/or light stress to encourage them to produce spores. These spores are then grown in balloons or photobioreactors to reach the sporophyte or gametophyte stage, for mass production. During this phase,

culture conditions are rigorously controlled, including illumination, bubbling, temperature and mineral nutrition. It is at this stage that microscopic individuals are brought into contact with cords for attachment. Some species of macro-algae have a specific organ, called a crampon, which enables them to attach themselves firmly.
Once the cords have been seeded, they are then wound around more robust ropes, which are installed at sea to enable seaweed cultivation. Cultivation can be carried out on ropes arranged horizontally or vertically, as required. These ropes can be attached to anchor points using a variety of methods, suspended from floating structures or fixed to the seabed using piles or dead bodies. However, rough seas can sometimes damage these structures or pull seaweed off the lines.
Each cultivation system requires a specific method for harvesting biomass. Most often, a vessel specially designed for the task is used. This vessel cuts the seaweed and brings it aboard for harvesting. To facilitate transport of the seaweed to land for processing or use, the vessel can be fitted with a filter press or centrifugal dewatering device, reducing the water content of the harvest from 90% to 20%.

6. Production control :

Seaweed production in the marine environment is controlled in an open, natural environment. In France, the use of fertilizers or phytosanitary products is not an option, as their dissemination in the environment cannot be controlled.

Consequently, seaweed cultivation relies on exploiting the nutrients naturally present in the environment, particularly near natural outlets such as estuaries, fish farms or shellfish farms. Incident light, essential for photosynthesis, is also an uncontrollable natural factor, depending on the location of the cultivation site, latitude, water turbidity, depth (up to a maximum of 30 meters) and tidal range.
Other factors such as temperature, seasonality and water circulation, linked to climatic conditions and ocean currents, also play a crucial role in algae cultivation. Certain macro-algae species, with specific requirements in terms of nutrients, salinity, temperature, light, depth and currents, are particularly well suited to large-scale

cultivation. However, a number of factors can affect cultivation, including predation, epiphytic plant growth, bacterial and viral infections, and pollution.

Macroalgae also offer interesting potential for the bioremediation of excess nutrients in the environment, which can lead to dystrophic crises. The cultivation of algae on eutrophic sites can therefore be particularly relevant. It's important to note that algae dry matter generally contains 7-8% nitrogen and 1% phosphorus, helping to eliminate these excess elements in nutrient- overloaded environments.

Moreover, the hatchery culture stage requires strict production control. All culture parameters and conditions must be rigorously controlled to ensure a successful transition to sea culture.

7. Seaweed cultivation in France :

France has recognized expertise in aquaculture techniques for several macro-algae species. In the 1990s, the CEVA (Centre d'Etude et de Valorisation des Algues) and Ifremer developed specific seaweed cultivation techniques, illustrated by technological demonstrations, notably with the *Undaria pinnatifida* species. These techniques were then adopted by micro-enterprises, mainly based in Brittany, who continued the artisanal production of macro-algae for food and cosmetic applications. Although the technical feasibility of cultivating this species was proven, it did not prove profitable on a large scale at the time, which limited mass production. However, this has encouraged the development of small businesses and a few seaweed growers in Brittany, who produce small volumes of seaweed for high value-added markets such as food and cosmetics. As a result, seaweed production in France today remains predominantly fishery-based.

However, seaweed farming is now back in the spotlight in France, with the emergence of new companies in this sector. Trials involving the co-culture of oysters and seaweed, as well as mussels and seaweed, are currently underway all along the Breton coast, as part of the CHACO project. This innovative approach makes it possible to enhance existing concession areas without affecting initial shellfish production, offering a promising prospect for the sustainable development of seaweed cultivation.

CHAPTER 4: INNOVATIONS IN ALGAE

1. New cultivation concepts :

In France, interest in the co-cultivation or co-breeding of algae and marine species such as fish, shellfish or shrimp is growing significantly. This concept, already present in various parts of the world through pilot experiments or operating aquaculture farms, offers promising prospects. In Mexico, for example, green algae from the *Ulva clathrata* species are cultivated in shrimp farming ponds. This co-production of seaweed and shrimp has several advantages: it improves the sanitary quality of the farm, reduces feed intake while maintaining equivalent growth, as the shrimp feed partly on the seaweed, and promotes better pigmentation of the shells and flesh of the shrimp, as the seaweed consumed is rich in carotenoids. In France, these practices are just beginning to be explored. A notable example of integrated multitrophic aquaculture, developed in Canada by Thierry Chopin, professor at the University of New Brunswick, illustrates these advances.

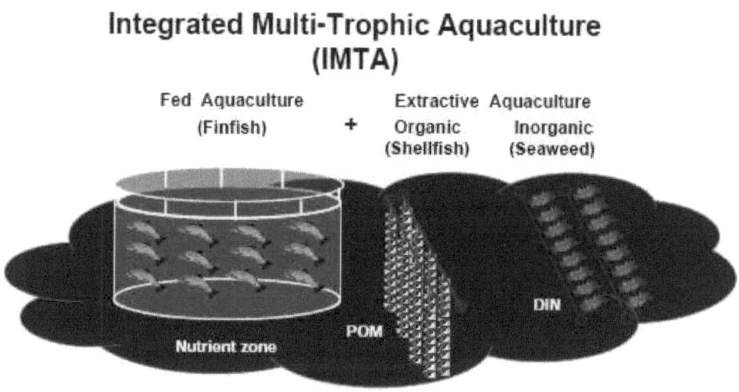

Figure 40: Conceptual diagram of an integrated multi-trophic aquaculture operation.

This conceptual diagram of an integrated multi-trophic aquaculture (IMTA) operation illustrates the synergy between feed aquaculture, centered on fish, and organic extractive aquaculture, focused on shellfish, which benefit from enrichment with particulate organic matter (POM). At the same time, inorganic extractive aquaculture, represented by algae, benefits from enrichment with dissolved inorganic nutrients (DIN), optimizing the efficiency of the entire aquaculture system.

2. Off-shore culture:

The creation of large offshore wind farms in France offers a promising development opportunity for aquaculture. This offshore wind farm deployment plan will create maritime zones where navigation and fishing will be regulated, and where the maintenance of the farms will require a human presence. This configuration paves the way for potential synergies with aquaculture activities, particularly macro-algae farming. The economics, as well as the environmental and societal impacts of this approach, merit in-depth study. For floating seaweed species such as *Sargassum*, floating culture, combined with structures to keep the seaweed within a defined area, could generate significant savings compared with traditional line-based systems. In experiments in the North Sea, ring-shaped structures, from 5 to 15 meters in diameter, have shown the best results in terms of productivity, although they do pose certain challenges during harvesting. A notable example of a candidate species for offshore aquaculture in the Pacific is the giant brown seaweed, *Macrocystis pyrifera*, renowned for its high light absorption capacity and rapid growth, doubling in weight every six months. However, deep-sea tests carried out by American teams revealed a number of difficulties, including access, mooring, nutrient supply and harvesting. The surface productivity of offshore crops, without inputs, varies between 20 and 50 tonnes of dry matter per hectare per year, depending on the devices and species used.

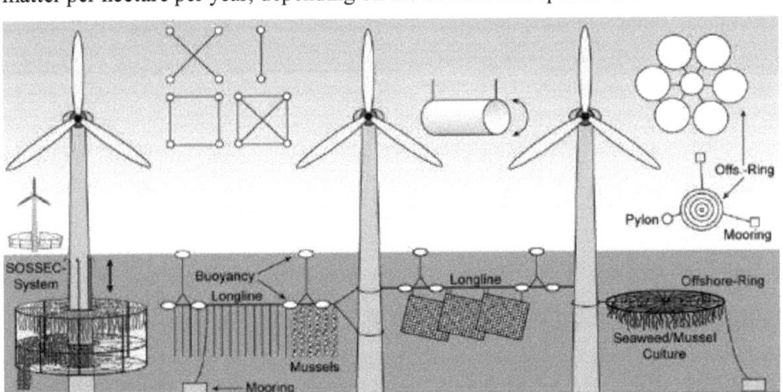

Figure 41: The multiple uses of offshore wind farms for seaweed and seafood co-culture.

3. Cultivation in bioreactors :

Cultivation in bioreactors, whether standard or photobioreactors (PBR), is very similar to microalgae cultivation, but is applied to macroalgae. It is based on the use of microscopic gametophytes, fragments of undifferentiated thalli, or micro-plantlets, to develop cultures of specific cells or tissues. This biomass, produced in a completely controlled manner, has applications in the cosmetics, nutrition and pharmaceutical sectors. For example, the production of rhodophyceae micro-plants enables the synthesis of high value-added molecules such as halogenated terpenes. The design of PBRs takes into account various parameters, such as light and energy supply, mass and heat transfer, as well as the choice of a reusable or single-use system. Total control of the culture in these environments enables strategies to be implemented to stimulate the synthesis of secondary metabolites, for example through light stimulation or the addition or omission of chemicals.

By using bioreactors or simple growing basins under natural sunlight with air, CO_2 and nutrient injection, it is possible to achieve surface productivities of up to 130 tonnes of dry matter per hectare per year. These results have been demonstrated in long-term experiments, such as the cultivation of *Gracilaria sp.* over a period of three consecutive years.

3.1 Varietal selection :

In Japan, the first seaweed growers, who relied on collecting *Laminaria* or *Porphyra* seeds, probably couldn't select specific strains of seaweed until they had mastered the reproductive cycles of these species during the 20th century. They were therefore entirely dependent on local genetic resources and the propagation success of propagules collected from culture media. It was only in the hatcheries developed in the second half of the 20th century that methods inherited from land-based agriculture, such as mass selection, enabled the domestication and improvement of some species in Asia. These advances have led, for example, to spectacular results in China with the selection of *Laminaria* cultivars accumulating maximum iodine.

Until the last decade, progress in genetics was virtually non-existent for marine algae. However, recent developments in marine genomics have given several macro-algae species access to high-performance tools and methods used for the genetic improvement of terrestrial plants. The search for molecular markers now contributes to a better understanding of the genetics of important traits, optimizing the efficiency of breeding programs. These markers are invaluable for rapidly testing varieties and selecting those with the desired characteristics. Marker-assisted selection is therefore crucial for identifying agronomically interesting traits.

Obtaining new varieties and strains of macro-algae, such as kelp hybrids, strains with high nutritional quality and resistance to disease and environmental stress, or varieties with high yields of molecules of interest or high energy potential, must become a central objective of future algal farming development projects.

The expected results should lead to the creation of a bio-bank of strains, housed within the French component of the European Biological Resource Centre EMBRC-Fr, comparable to a catalog of varieties, the accessibility of which will have to be precisely defined for researchers and industrialists. Moreover, this technological know-how in variety creation could become an international challenge, opening the way for existing or new companies to disseminate these innovations in algae-producing countries, particularly in Asia.

3.2 Comparison of sales prices :

Table 8: Comparison of macro-algae selling prices.

Growing/harvesting techniques	Price list

Scoubidou fishing - Fresh seaweed (*Laminaria*)	40 - 50 euros per tonne
Hand-picked - Fresh seaweed (10 to 20 species)	35 - 2,000 euros per tonne
Stabilized algae (Micronized flours and powders)	1,000 - 15,000 euros per tonne
Growing in lagoons (Hypothesis: green or red algae)	500 - 800 euros per tonne of dry matter (50 euros for optimistic estimates)
Offshore culture (Hypothesis: Brown algae)	200 - 315 euros per tonne of dry matter
Offshore culture (Edible brown algae)	500 - 1,200 euros per tonne of fresh material

Note: The wide variation in sales prices for each technique is explained by the diversity of seaweed species, the volumes harvested, and above all by the destination for the seaweed.

4. Pretreatment and extraction processes :

4.1 Drying - Stabilization :

Once harvested, macro-algae can be used directly, in particular for alginate extraction. In this case, it is crucial that the processing sites are located close to the harvesting sites. Using wet algal biomass, fresh from the sea, preserves the original properties of the algae, protecting sensitive molecules, while ensuring higher yields and purity. However, this requires rigorous logistical management.

The high water content of algal biomass poses a problem when it is not used immediately, due to the risk of fermentation. Consequently, a drying and stabilization step is essential to avoid degradation and to transport algae to industrial sites without compromising quality. Several methods can be employed.

On-site pre-banning is the method most commonly used in major producing countries such as Asia and South America. This technique involves leaving the algae to dry in the sun, sometimes under cover, with the wind as an auxiliary. The water content is thus reduced from 80% to around 30%. This process is particularly suited to the climates of these regions, where it reduces costs.
In Europe, industrial drying is more widespread. It uses kilns and dryers to stabilize the seaweed and transform it into flour, with a water content of less than 10%. A gentle drying process, which preserves sensitive molecules without denaturing them, has recently been developed by the Aléor company.
Another, more economical method of desiccation is to mix the algae with absorbent materials. Straw, for example, can act as a dehydration aid and structuring agent, preventing anaerobic putrefaction and orienting fermentation towards aerobiosis. This process can be used to produce organic soil improvers such as compost. Drying algae remains a key stage in extraction, fractionation and biorefinery processes. Excessive drying can lead to high costs, alter algae properties and complicate separation of constituents.

4.2 Grinding :

Grinding is an essential operation to facilitate packaging and transport of the material, as well as to improve yields during solid-liquid extraction, where the extractability of certain compounds may be limited by diffusion. Seaweed can be chopped or ground in either a wet or dry state. For wet grinding, dilution pumps and homogenizers are particularly suitable. Food industry mincers can also be used to reduce seaweed to fragments of just a few millimeters. Wet milling can be assisted by cold (cryo-grinding), a technique that delivers very interesting results in terms of extraction efficiency and preservation of active ingredients. This process is used in certain industries to manufacture seaweed pastes and extracts, valued for their phytostimulant properties.
Grate mills equipped with knives or rotary hammers are the most commonly used mills for grinding dry materials. The addition of cold, in the form of liquid nitrogen, improves grinding efficiency by taking advantage of residual moisture (around 10%

of dry matter), and preserves heat-sensitive active molecules by limiting heating due to dissipated mechanical energy.

Coarse grinding produces cut sizes of 0.5 mm, 1 mm, 2 mm or 5 mm, while fine grinding produces cut sizes in the 100 to 200 micron range, or even below 100 microns.

5. Extraction of compounds of interest :

5.1 Extraction and purification methods for compounds of interest :

Obtaining concentrated and isolated molecules of interest involves several steps specific to each type of molecule.

5.1.1 Solvents :

Various solvent extraction methods can be applied to macro-algal biomass.

Water is the solvent of choice for extracting molecules such as colloids (alginates, carrageenans, ulvans, starch) and minerals. Hydroalcoholic mixtures are used to extract polyphenols. For the extraction of pigments, such as carotenoids, apolar solvents or supercritical CO_2 are used, thus avoiding the use of organic solvents. Astaxanthin, for example, is extracted using this technology. However, organic solvent extraction is still complex to implement on an industrial scale.

5.1.2 Purification :

Extracts are also purified by a variety of methods. For colloids, the precipitation/solubilization method is commonly used. Ultrafiltration or chromatography methods are used to produce molecules with very high added value. Ultrafiltration is used in the cosmetics and food sectors, while chromatography is favored by the pharmaceutical and medical industries.

5.1.3 Biotechnology: hydrolysis of macromolecules :

Once extracted and purified, macromolecules can be hydrolyzed to obtain others, smaller in size but of greater interest. For example, the depolymerization of a polysaccharide into oligosaccharides by hydrolysis makes it possible to obtain active ingredients with better diffusion and bioavailability, thus increasing their biological activity. Hydrolysis can be carried out enzymatically, using enzymes obtained from marine bacteria, or by fermentation, also involving marine bacteria or other micro-organisms.

The use of micro-organisms derived directly from the marine environment is explained by their adaptation to the transformation of algal matter, in particular for the hydrolysis of polysaccharides specific to algae.

Enzymes and bacteria with hydrolytic activity towards different seaweed polysaccharides (alginates, carrageenans, ulvans, laminaranes, agar-agar, porphyran, etc.) have been identified and are the subject of research worldwide.

By way of example, the production of oligo-alginate for skin cosmetics is based on this technology. This technology is particularly interesting for large-scale applications in the energy or chemical sectors. It makes the sugars and uronic acids that make up polysaccharides available for further processing, such as acetobutanolic or ethanolic fermentation. The use of enzymes or bacteria also improves extraction yields of compounds of interest by facilitating the destructuring of algal walls and fractionation of extracts.

5.2 Agar extraction :

In the traditional process, dried seaweed is chopped and then boiled in water to which an acidic ingredient, such as acetic acid or dilute sulfuric acid, has been added. The resulting hot liquid is filtered to remove residual seaweed pieces, then neutralized with an alkaline agent, such as sodium bicarbonate. A jelly forms when the temperature drops below 40°C. To eliminate the seaweed taste and light coloration of this jelly, it is subjected to a process of freezing, thawing and dehydration, repeated two or three times to progressively purify the agarose.

5.3 Sterol extraction :

Sterols are extracted from the total lipid fraction of the algae using solvents. Green" solvents, such as EtOH/H2O-80/20, are preferred as they are less expensive, less toxic and more environmentally friendly than traditional solvents such as hexane or acetone. Sterols are then isolated from the total lipid fraction by chemical (saponification) or enzymatic (lipase) methods. Enzymatic hydrolysis offers higher yields, ranging from 3 to 6%, compared with only 1 to 2% for saponification.

5.4 Alginate extraction :

After harvesting, the seaweed is cut, cleaned and then macerated in dilute mineral acid for demineralization. The alginate present in the seaweed is transformed into insoluble alginic acid, and undesirable constituents such as fucoidan, mannitol and mineral salts are removed. The demineralized seaweed is then ground in the presence of an alkali or alkaline salt to neutralize the alginic acid and solubilize it in the form of a salt corresponding to the alkali used. Insoluble compounds are removed by filtration, flotation or decantation. The alginate is then precipitated, pressed, washed, dewatered, neutralized, dried and ground to the desired particle size. The salts obtained have specific properties: alkali metal alginates (sodium, potassium, magnesium) are soluble in water, while alginic acid and its calcium derivatives (calcium, aluminum, iron) are practically insoluble. Soluble salts dissolve at low temperatures to form thick, viscous solutions, producing heat-resistant gels when cross-linked by divalent cations such as calcium or trivalent cations such as strontium.

5.5 Carrageenan extraction :

Carrageenan extraction varies according to the purity required. To obtain pure carrageenan, a so-called fine extraction is required. However, a simple treatment with potash can produce semi-refined carrageenan, which is also authorized as a food additive in the same way as refined carrageenan.

5.5.1 The case of refined carrageenans :

The extraction process for refined carrageenan, as illustrated in figure 48, follows a fine method widely used in the industry. Carrageenan, especially when it has been dried for a long time, is first soaked (stage 1) in a calcium chloride solution and shaken vigorously to remove sand, shells, larvae, epiphytes and other impurities.

The calcium then penetrates the tissues, fixing the carrageenan in the cell walls to prevent loss during rinsing (stage 2).

During this rehydration phase, a preliminary extraction on a representative sample of the biomass is recommended to determine the optimum parameters for obtaining the best yield. The thallus is then immersed (stage 3) for two hours in an alkaline solution (NaOH, 1 N) at 65°C, at a rate of 100 liters per kilogram of dry tissue. Under the effect of a sledgehammer, the plants gradually disintegrate into a colored magma (stage 4), dissolving the carrageenans in the form of sodium salts. It is essential to maintain a basic environment throughout the process, as carrageenans tend to depolymerize in an acidic environment at high temperatures. Separating the carrageenan-containing liquid phase from the other materials is a delicate stage. Filtration (stage 5) takes place in a cylinder kept under pressure and at high temperature, producing a yellowish liquid containing 2 g of carrageenans for 98 g of water. The challenge then lies in recovering these 2 g of carrageenan.

To achieve this, the liquid is sprayed into a tank containing isopropyl alcohol (stage 6); this causes the carrageenan to flocculate into whitish spongy clusters, stirred by the blades of a large propeller. Another kappa-carrageenan precipitation process relies on specific interaction with potassium salts, eliminating the need for an organic solvent such as isopropanol.

Once this operation is complete, a powerful pump sucks up the "carrageenan-alcohol precipitate" suspension and deposits it on a horizontally vibrating metal screen (stage

7), which retains the spongy clusters while allowing the alcohol to pass through. The carrageenans are then rinsed a final time with pure isopropyl alcohol.
At this stage, the extract is still fragile and subject to depolymerization. It must therefore be dried at low temperature (5°C) under vacuum, in an inverted "bell jar". Although time-consuming, this process produces highly polymerized carrageenans. Some manufacturers, however, prefer to speed up the process by entrusting part of the dehydration to a centrifuge, although this can be detrimental to the quality of the final product. The 85% of the isopropyl alcohol used in these operations is recovered by a distillation column. The dry fibrous mass is then ground into particles ranging in size from 80 µm to 270 µm, depending on requirements.

After this stage, rigorous checks are carried out to determine the product's exact chemical composition and functional properties, such as residual moisture, purity, hot and cold viscosity, and gel strength in milk and water with KCl. If the type of carrageenan is to be used in agri-food products, it must also meet very strict specifications, including analyses of trace metal elements and microbiological quality tests.

5.5.2 The case of semi-refined carrageenans :

Approved for use in food products under code E407a, semi-refined carrageenans are known by a variety of names, including "Alkaline Carragheen Flour" (ACF), "Philippine Natural Grade" (PNG27), "Processed Euchema Seaweed" (PES), and "Semi-Refined Carrageenan" (SRC). The principle behind their production is based on the treatment of biomass with an 8.5% concentrated potash solution at a temperature of 60°C, where the potassium immobilizes the polymer inside the plant tissue. The biomass is then dried and ground into a powder. PNGs have similar qualities to pure carrageenans, notably in terms of viscosity and gel strength. The procedure for obtaining PNGs is faster and less costly than the fine extraction of refined carrageenans. What's more, this method can be carried out directly at the carrageenan harvesting sites, considerably simplifying the logistical process. Current worldwide production is estimated at around 30,000 tonnes of refined carrageenan (E407 additive) and 18,000 tonnes of semi-refined carrageenan (E407a seaweed flour).

5.6 Anaerobic digestion of macro-algae :

Anaerobic digestion can be applied both to whole algae and to extraction residues, such as those from alginate or carrageenan production. This process is based on the same principles as conventional anaerobic digestion, enabling both the mineralization of organic matter, the recovery of energy in the form of methane (CH_4) contained in the biogas resulting from digestion, and the recovery of mineral nutrients such as nitrates (NH_4^+) and phosphates (PO_4^{3-}).

Although based on an ancient concept, methanization has proved its effectiveness and robustness. Today, it is widely used, notably in the management of agri-food effluents and sewage sludge. The methanogenic potential of algae is comparable to that of more conventional substrates, such as slurry centrifuge rejects or cattle manure, with a yield of 200 to 300 Nm^3 of biogas per tonne of volatile matter introduced.

However, the methanization of marine macro-algae presents several technical challenges linked to their composition. These challenges include a high content of non-digestible mineral fraction, the presence of methanogen inhibitors such as sodium and chlorides, as well as the presence of sulfur, which is found in the biogas in the form of H_2S.

Co-digestion, an approach combining macro-algae with other substrates, is also conceivable. For example, sewage sludge or the fermentable fraction of household waste could be added to the algal biomass, which could optimize the methanization process and improve yields.

6. Applications, potential markets :

6.1 Introduction :

Worldwide, seaweed is mainly considered a vegetable, with three quarters of macro-algae production destined for direct human consumption, playing a central role in traditional Asian cuisine. By contrast, in Europe and North America, the industrial uses of macro-algae are more diversified and well established, notably in the colloid industry. Other traditional applications, such as agriculture and cosmetics, are also widespread.

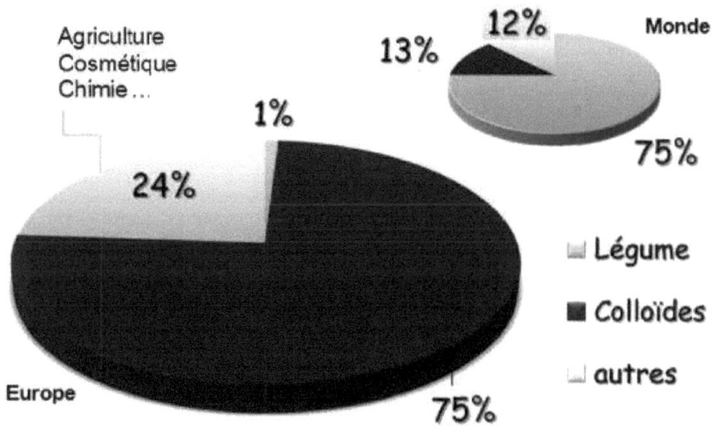

Figure 42: European and world markets: application families.

Marine plants, in whole or fractionated form, are ubiquitous in our daily lives, and have a wide variety of uses. In addition to their use as food, seaweed is playing an increasingly important role in sectors such as agriculture, where it is used as animal feed and organic fertilizer, and in the pharmaceutical industry, where it is used to coat drugs or is appreciated for its slimming properties. They are also used in cosmetics (soaps, creams, toothpastes), cleaning products and textiles. With the rapid expansion of the agri-food industry, seaweed is now inextricably linked with the notion of food. For example, the average European unknowingly consumes seaweed-derived texture agents at least five times a day, in a variety of prepared foods. Some seaweed extracts, with their gelling, thickening or texturizing properties, are comparable to substances found in terrestrial plants, such as starch and pectin, or in animal marrow, such as gelatin. Following the mad cow crisis in 1996, the food industry increasingly turned to plant-based alternatives with properties similar to gelatin, such as phycocolloids.

6.2 Phycocolloids :

Since the beginning of the 20th century, macro-algae have been exploited for the production of phycocolloids, the most commonly used of which are alginates, carrageenans and agars. These polysaccharides, which are not found in terrestrial plants, are the main industrial products derived from algae. Although some polymers with similar properties can be produced by terrestrial plants, such as gum arabic, phycocolloids have unique characteristics as texturizing agents. Not all phycocolloids have the same functional properties: agars and Lambda carrageenans are mainly used as thickeners, while Iota and Kappa carrageenans are recognized for their gelling capacities.

Alginates, on the other hand, are versatile, acting as both thickeners and gelling agents. These properties are widely exploited in a variety of sectors, including food and feed, pharmaceuticals, agriculture and even medicine.

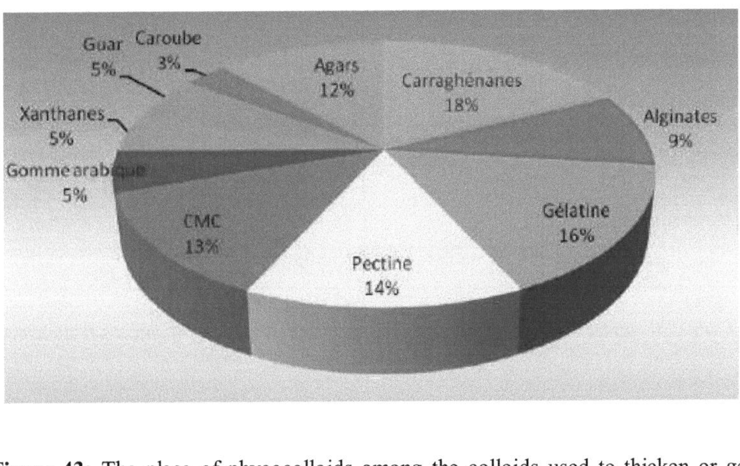

Figure 43: The place of phycocolloids among the colloids used to thicken or gel aqueous solutions.

Altogether, these phycocolloids account for over 39% of the colloids produced worldwide, with a global market of around 600 million US dollars a year.

These products are referred to as "demographic" because of their widespread use in the preparation of food and convenience products. Demand for these phycocolloids is growing in parallel with population growth, changing eating habits marked by the development of the agri-food industry and the growing consumption of processed products, as well as in the wake of health crises, such as the suspicion surrounding bovine gelatin generated by bovine spongiform encephalopathy (commonly known as "mad cow disease").

Table 9: Overall annual value of algae products.

SEAWEED PRODUCTS	TOTAL ANNUAL VALUE
Human Food (Nori, aonori, Konbu, wakame, etc.)	$5 billion
Agar (Food ingredient, pharmaceutical, biological/microbiological)	132 million
Alginate (Textile printing, food additive, pharmaceutical, medical)	$213 million
Carrageenan (Food additive, pet food, toothpaste)	$240 million
Other uses of seaweeds	
Fertilizers and conditioners	$5 million
Animal Feed	$5 million
Macroalgal Biofuels	Negligible
TOTAL	$5.5-6 billion

Alginates are polymers found in the cell walls of a wide variety of brown seaweed species. They are commonly used as gelling and/or thickening agents in a variety of industrial fields. In the food and pharmaceutical industries, alginates are used as emulsion and suspension stabilizers. For example, they are found in products such as ice cream, jam, creams, custards, lotions, toothpaste, and as coatings for pills.

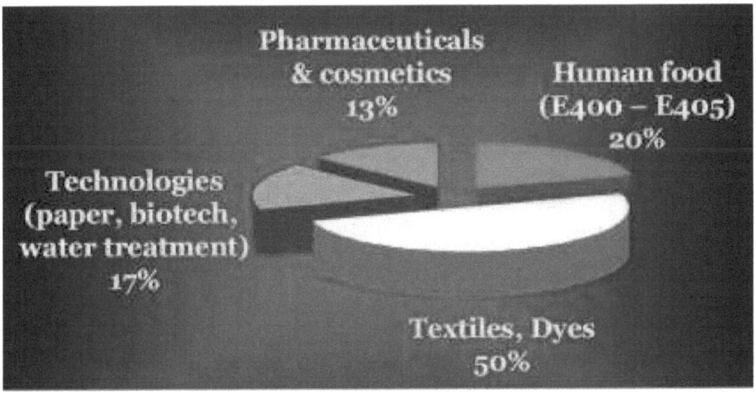

Figure 44: Application markets for alginates.

Alginates are also used in the production of paint, certain building materials, glue and paper, as well as in the oil, photographic and textile industries. Their ability to bind heavy cations makes them valuable as depollutants and detoxifiers. The majority of brown algae used for alginate production are harvested from the wild rather than cultivated specifically for the purpose. However, in China, some kelp crops, originally destined for human consumption but now decommissioned, are reused as raw material for alginate extraction. Overall, the value of the alginate market was estimated at around US$213 million in 2003.

Carrageenans are polymers derived from the cell walls of red algae, to which one (κ-), two (ι-) or three (λ-) sulfate groups are attached. Thanks to their specific properties, carrageenans, known by their industrial name E407, are widely used in human foods. They are found in products such as milk and milk derivatives, creams, ice creams, cakes, custards, mousses, gelled yoghurts and sauces. These phycocolloids give these foods a texture, cohesion and consistency that prevent them from liquefying.

Carrageenans are also used in animal feed, notably in the form of kibbles mixed with other feeds, and their use is increasingly common in cosmetic formulations.

The majority of seaweed used for carrageenan production is cultivated, as demand for raw materials exceeds what natural resources can offer. In 2003, the value of the carrageenan market was estimated at around US$240 million.

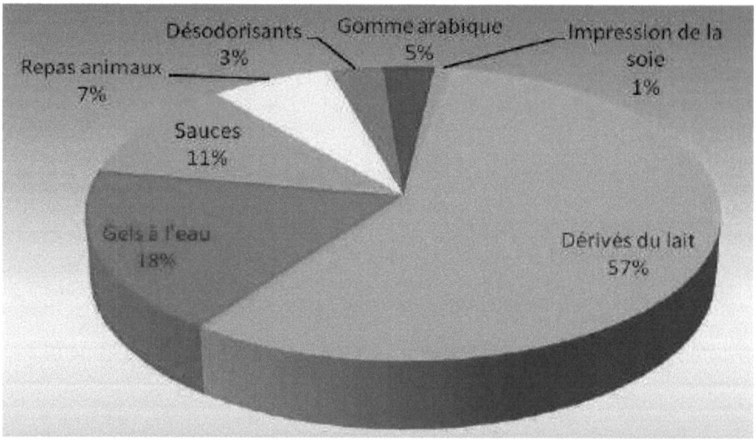

Figure 45: Distribution of the main uses of carrageenans.

Agars are polymers derived from the cell walls of red algae. The algae used for commercial agar production are either harvested from the wild or cultivated. Like carrageenans, agars are used as stabilizers for emulsions and suspensions, and as gelling and coating agents for foods. Around 90% of agar production is destined for food applications, while the remaining 10% is used for bacteriological and other biotechnological purposes. In 2001, agar production totaled some 7,630 tonnes, with an estimated market value of around US$132 million.

Table 10: Current uses of κ, ι and λ carrageenans

Compound	**Effect**	**Application**
κ-carrageenan	Gelling agent	Chocolate milk Ice creams and dessert creams Canned meat

		Pet food Air freshener gels Enzyme immobilization *In vitro* cultures
ι-carrageenan	Gelling agent	Desserts Sauces Ice creams and ice cream Cosmetics
λ-carrageenan	Thickener	Desserts Sauces Toothpaste Cosmetics

6.3 Human food :

Macro-algae play an important role in various sectors of human nutrition. They are mainly consumed as vegetables, particularly in Asian countries, where this sector accounts for 75% of global production. Their nutritional qualities also position them in the dietary supplements market, while their high polysaccharide content makes them valuable in the food colloid industry.

6.4 A food in its own right :

In Asia, seaweed is a staple of the traditional diet. It is used daily in salads, soups, condiments and as a vegetable side dish. In Japan, for example, each inhabitant consumes an average of 1.3 kg of dried seaweed per year, or 3.6 g per day. Although the quantities consumed are not very high, daily consumption of seaweed is an integral part of the Japanese diet, with proven health benefits.

This ancestral know-how has enabled the Japanese to make the most of each seaweed and its unique properties. Each type of seaweed is prepared differently to enhance its color, texture or taste. For example, the rich sodium glutamate content of kelp is exploited to prepare Miso soup, while the flavor of Nori is sublimated in dried or toasted Nori leaves, often eaten for breakfast with rice, in maki or onigiri. The taste and color of Nori leaves vary according to their place of origin, with high-quality Nori leaves - totally black, smooth and shiny - often given as gifts at weddings. Wakame seaweed is generally eaten in salads, condiments or soups, where its bright green color and crunchy texture are highlighted. Konbu is often served as a vinegar condiment, while agar-agar, derived from *Gelidium* or *Gracilaria*, is widely used to prepare all kinds of flavored jellies served as desserts.

In France, on the other hand, seaweed doesn't generate the same enthusiasm. The term "algue" does not arouse the enthusiasm of most consumers. And yet, because of their nutritional value, these "sea vegetables" deserve to be more widely integrated into Western diets.

The nutritional value of macro-algae lies mainly in the combined presence of three major components: fiber, minerals and protein. Added to these are metabolites with antioxidant and anti-free radical properties, such as carotenoids, polyphenols, vitamins and polyunsaturated fatty acids. A detailed description of all the components present in macro-algae is given in the "biochemical composition" section.

6.5 A potential source of food supplements :

An epidemiological study carried out in Asia (Teas, 1981; Carper, 1989) showed that regular consumption of seaweed could be linked to better health, notably by reducing the risk of colon, breast and prostate cancer.

The diversity of molecules making up macro-algae gives them multiple nutritional qualities, making them ideal for use as food supplements, particularly in the slimming and health fields. The main avenues being explored are described below.

The fiber and sterols present in seaweed have been shown to reduce plasma cholesterol. For example, green and brown seaweed contain sterols closely related to phytosterols, which could make them a new source of cholesterol-lowering phytosterols. Sulfated seaweed polysaccharides have antitumor and immunostimulant properties. Metal chelation, made possible by alginates, plays a detoxifying role, as has been exploited in the treatment of French populations affected by lead poisoning. Halogenated compounds, such as iodine, have antiseptic and antibacterial properties, and influence lipid metabolism. The calcium contained in algae is highly bioavailable and recommended for preventing bone decalcification associated with menopause. Finally, secondary metabolites, such as polyphenols and pigments, prevent the deposition of atherosclerotic plaques in blood vessels and play a key role as antioxidants against oxidative stress, responsible for cellular aging.

6.6A potential source of texturizers :

Although the French are not fond of seaweed as a vegetable, they consume it on a daily basis, often without realizing it. Indeed, seaweed extracts are commonly used as texturizing additives (gelling agents, thickeners, stabilizers) in products such as creams, custards, soups and cold meats. The main use of seaweed is in the colloid industry: alginate, agar and carrageenan. The diversity of properties of these substances leads to a wide range of applications.

These phycocolloids can be used as stabilizers or emulsifiers in various foods such as salad dressings and ice creams, but they also act as film-forming agents, protecting against bacteria, desiccation, or damage due to various handling.

Alginates, extracted from brown kelp, are widely used in the food industry for their ability to gel and form highly viscous solutions. They can act as thickeners, gelling agents, stabilizers and cryoprotectants for frozen foods, or as edible protective films. They are found in ice creams and sorbets, where they control ice formation, act as flavor enhancers, and provide a soft, pleasant texture in the mouth. They are also present in whipped cream, rice pudding, whipped yoghurt and restructured fruit, vegetable, meat and fish purées. Chemically modified by hydroxypropylation, they stabilize emulsions in mayonnaises and salad dressings. They are also used in the breading of breaded fish and in the clarification of liquids such as oils, wine, vinegar and beer. When used in the European Union, they are listed as ingredients under codes E400 to E405, depending on the form of the alginate.

Agar, extracted from red seaweed, is used in a wide range of applications in the food industry. In particular, it is used as a stabilizer for fruit jellies, cream, flavored milk, ice cream and sorbets. It is also found in other food products such as frozen foods, icings, meringues, and gel-form foods like desserts, candies and fruit juices. When used in the European Union, agar is listed under the code E406.

Carrageenans are a group of water-soluble polysaccharides more widely used than agar as gelling and thickening agents in many foods, especially milk-based products. They are present in chocolate milk, ice cream, evaporated milk, puddings, jellies, jams, salad dressings, gelatin desserts, meat products, and ready-to-prepare powdered custards. When used in the European Union, they are listed under codes E407 for refined carrageenans and E407a for semi-refined carrageenans.
There are many other phycocolloids, such as ulvans, xylans, rhamnans, hemicelluloses, celluloses, mannans and arabinogalactans, most of which remain unexploited to this day.

7. Health - Pharmaceuticals :

7.1 Many potential molecules, few commercial applications :

The molecules most commonly used in the health and pharmaceutical sectors are those already exploited in the food industry, namely alginates, carrageenans and agar. However, macro-algae contain a much wider range of molecules of pharmaceutical interest. Although many of these molecules are still at the research stage or are of potential interest, very few are currently marketed in this sector. Here is a brief description of their commercial applications and potential action.

7.2 Alginates :

Alginates are widely used in the pharmaceutical industry for their ability to chelate and form highly viscous solutions. They are used in dental impression pastes, and as antacids to treat heartburn and gastric reflux. In the medical field, alginates are used to treat wounds, notably in hemostatic compresses and hydrocolloid dressings for burns. They are also used in the composition of artificial livers.

7.3 Agars :

Agars are used in the medical and pharmaceutical fields to produce swelling agents, laxatives, suppositories, capsules, tablets and anticoagulants.
They are also indispensable in the manufacture of biological culture media, and play a crucial role in molecular biology, particularly for separation methods such as electrophoresis.

7.4 Carrageenans :

Carrageenans offer many potential uses as pharmaceuticals. They have antitumor, antiviral and anticoagulant properties, and immunomodulatory activities have also been explored.

7.5 Sulfated polysaccharides :

Sulfated polysaccharides, such as carrageenans, fucans and ulvans, are of particular interest in the pharmaceutical field. The main research focuses on fucans and fucoidans, ulvans, as well as other oligosaccharides and specific molecules.

7.6 Fucans and fucoidans :

Fucans and fucoidans, extracted from brown algae, are complex polysaccharides of the poly-fucose family with variable sulfation. They have a wide range of properties. They could, for example, prevent and/or relieve gastric ulcers, prevent associated gastric cancers, and avoid liver disorders. Products of this type already exist on the Japanese market. Fucans also have antioxidant and immunostimulant properties. Their similarity to heparin gives them antithrombotic and anticoagulant properties. Their anti-cancer, anti-angiogenic and anti-tumor properties are the subject of much research into the treatment of various cancers. They also have bacterial anti-adhesion and even bactericidal properties, as well as antiviral activities.

To maximize these properties, it may be necessary to reduce the molar mass of these molecules by depolymerization, which can be carried out chemically (acids, radicals) or biochemically (specific enzymes).

7.7 Ulvanes :

Ulvans, sulfated polysaccharides extracted from green algae, belong to the family of sulfated rhamno-aldobiuronans. Their properties are comparable to some of those of fucans, but they also have others, as complex polysaccharides composed of rare sugars. Although less studied, their pharmaceutical potential remains promising.

Other sulfated polysaccharides derived from macro-algae exist and present a "diversity to be investigated". Certain oligosaccharides, obtained by depolymerization, may also offer potential action in the pharmaceutical sector. These oligosaccharides are produced by enzymatic hydrolysis using marine bacteria, making it possible to process biomass of similar origin.

In addition, other specific molecules derived from macro-algae are being considered for pharmaceutical applications. These include microcrystalline cellulose, floridosides and terpenes. Griffithsin, an antiviral glycoprotein isolated in 2005 from a New Zealand red alga of the Ceramiales family, *Griffithsia sp.*, is the subject of intensive research as a vaginal microbicide to limit the transmission of AIDS. In addition, the development of *Caloglossa leprieurii* algae cultivation could provide a new source of an antihelmintic, a drug used to expel parasitic worms.

8. Cosmetics :

Macro-algae are a key ingredient in marine cosmetics, recognized worldwide for their uniqueness and success. The use of macro-algae in the cosmetics sector is developing in several directions. Their excellent dermocompatibility positions them in the thalassotherapy market.

As in food, they play a texturizing role in the formulation of cosmetic products such as masks and toothpastes. In addition, they are a source of natural surfactants and active ingredients, enabling the production of detoxifying cosmetics with beneficial effects for the skin.

Oligo-alginates, derived from phycocolloids, have anti-pollution, anti-eruption, anti-acne and anti-aging properties. Fucoidanes help reduce skin aging and promote hair growth.

9. Agriculture :

9.1 Fertilizers :

Historically and traditionally, macro-algae have been used as fertilizers throughout the world, particularly in coastal regions. Sea lees" are spread on fields, improving the mineral composition of impoverished soils thanks to their mineral content, and increasing the water-holding capacity of soils thanks to their rich colloid content, which helps to bind silt and reduce erosion of arable land. Brown macro-algae are also used in the formulation of growing media (enriched potting soils) and fertilizers in the form of flours or liquid extracts.

Once the molecules of interest or the carbohydrates have been extracted from the algae, the residual biomass, rich in nutrients, can be used as a bio-fertilizer.

9.2 Phytosanitary treatments :

Macro-algae offer a range of beneficial properties in the crop protection sector. They contain compounds such as phytohormones (cytokinins, auxins, gibberellins) that promote germination, leaf and stem growth, and flowering. They can also be used as biological protection agents against plant diseases by acting as elicitors, i.e. stimulating the plant's natural defenses.

These plant "vaccines" help to reduce the frequency and importance of curative phytosanitary treatments based on fungicides or chemical pesticides. Compounds such as fucoidanes, laminaranes and oligo-alginates of the mannuronic polyacid type possess this type of activity.

These substances can be found in the form of flours activated with metals such as copper. They can also be used as formulation aids in the form of green surfactants (grafted algae polysaccharides) or anti-bounce polymers. Some companies market extracts titrated with oligosaccharide actives that stimulate plants' natural defenses, for use in foliar treatments and watering.

Products derived from macro-algae, approved as phytosanitary products and recognized by professionals, particularly arboriculturists, are already available on the French market. These products protect plants by stimulating their natural defenses.

9.3 Environmental benefits :

Another sustainable option is the production of certified organic fertilizers. When applied to crops, the nutrients contained in algae and algae extracts are released slowly, promoting plant growth. More importantly, chemical fertilizer production is energy-intensive and contributes significantly to greenhouse gas emissions. Given that demand for fertilizers is set to increase over the coming decades, the production of algae-based fertilizers offers significant potential for reducing the use of chemical fertilizers and mitigating their negative environmental impacts. Algae are particularly rich in major nutrients such as nitrogen and potassium, as well as secondary nutrients such as sulfur, calcium, magnesium and trace elements.

9.4 Animal feed :

9.4.1 Nutrition :

Once used traditionally in certain coastal regions, where livestock grazed on the shore, seaweed is now a common ingredient in animal feed for a variety of applications. They can be found in the form of flours in animal feed formulations. They can also be used as food supplements in animal nutrition. Algae are recognized for their prebiotic, antioxidant, immunostimulant and other effects, which are exploited in feeds for laying hens and young rabbits, for example, or in mineral salt stones (iodine, etc.) for cattle, sheep and goats.

In the colloid industry, carrageenans are used in pet food as thickeners or suspension stabilizers. Alginates are used to give granules their shape.

9.4.2 Aquaculture :

Macroalgae also have applications in aquaculture. For example, abalone farming, a booming industry in Chile and potentially in France, requires around 100 tonnes of fresh algae for every tonne of abalone produced. As harvesting from natural populations cannot support this demand, farmers must turn to the production of feed algae by algoculture. It is also possible to use enzymatic digestion of whole macro-algae to prepare feed suitable for the prey used to feed fry.

Green algae extracts can be used as palatability enhancers in salmon feed formulations, where fish oil is replaced by vegetable oil. In addition, some algae extracts have immunostimulant properties for various aquaculture species, such as turbot and shrimp. The aforementioned shrimp/green algae co-culture, practiced in Mexico, improves the sanitary quality of farming, reduces feed costs and improves the pink pigmentation of shellfish raised under these conditions.

9.5 Bio-sourced materials :

Extracts from macro-algae, such as alginates, are already used in various materials (yarns, films) for medical applications. The best-known example is calcium alginate

yarn, which is used to manufacture the haemostatic compresses commonly used in hospitals around the world. Calcium alginate fibers, obtained by wet spinning, also have excellent thermomechanical properties and high fire resistance, which has made them useful as fibers for fire protection garments.
There are also mixed cellulose/flour brown seaweed viscose fibers on the market, used in furnishing textiles, sportswear and underwear. Alginates are widely used in hydraulic-bonded materials (adhesives, cements and mortars) as rheology agents and setting regulators. They are also found as aids in the manufacture of ceramics and as impression materials in sculpture and dental prosthetics. They are also widely used in the paper and cardboard industry. It is possible to manufacture papers from algae fibers or wood pulp/algae fiber blends. Some products of this type are already marketed in the art and decorative paper segment. Cardboard and molded cellulose containers, such as egg cartons and horticultural pots, are also possible. For paper/cardboard applications, the absence of lignin in algae facilitates the use of celluloses/hemicelluloses. Algae colloids, such as alginates, are used as texture agents for dyes and paper sizing.
Recent research has shown that algae and their extracts can also be used as fillers or functional additives in thermoset or thermoplastic matrix composites. Developments are underway in the thermoplastics sector for packaging films and injected parts.

9.6 Bioenergy, biorefinery :

The feasibility of industrial co-production of mineral chemical intermediates (potash, iodine, chlorides, sulfides) and organic intermediates (acetone, acetic acid, butyric acid, butanol, methyl or ethyl acetate) by fermentation of marine macro-algae was demonstrated as early as the First World War in the United Kingdom and the United States.
At that time, with German chemistry having ceased all exports to these countries, a veritable chemical industry based on the exploitation of natural macro-algae deposits developed, notably in California with the exploitation of *Macrocystis pyrifera* fields. Algae biorefining then developed around the phycocolloid industry, which, all things considered, is the macro-algae industry's equivalent of the wood pulp industry. Algal

polysaccharides can be used to produce emulsifiers and surfactants using processes similar to those used for higher plant polysaccharides. Algal polyphenols are ideal bases for the formulation of glues and thermosetting resins.

In recent years, a new wave of R&D projects has focused on the large-scale production of macro-algae for conversion into biofuels by fermentation. There are several potential avenues, but some are more advanced than others. Most macro-algae have a low lipid content, which rules out the production of biodiesel by diester-type processes. However, it is possible to convert macro-algal biomass into oil by pyrolysis. Species rich in polyphenols, such as brown algae found in the tidal range, are particularly well-suited to this process. Macroalgae, rich in polysaccharides, can be fermented to produce bio-alcohols (ethanol, n-butanol, isobutanol) and bio-methane. Large algae contain both common polysaccharides, such as glucans (starch, laminarane, cellulose), whose transformation processes into bioethanol are well known, and more original polysaccharides (alginates, carrageenans, ulvans, agar-agar) for which R&D is still needed.

Demonstration projects for the fermentative conversion of brown macro-algae into isobutanol are underway in the United States, Chile and Norway. Similar projects to convert carrageenan into ethanol are underway in Korea, the Philippines and Japan. These initiatives involve consortia of chemical and fuel manufacturers, as well as academic research teams. Internationally, major chemical companies such as DuPont and Dow Chemical are planning to use algae to produce certain chemicals essential to their industry, such as ethanol and isobutanol.

The most advanced pre-treatment methods for producing bioethanol from algal polysaccharides are currently at the research and development stage.

The waste streams remaining after extraction can then be subjected to anaerobic digestion. Biogas production by anaerobic digestion of macro-algae is a highly promising method of producing gaseous biofuels. This technology eliminates several major obstacles that contribute to the high costs of algae biofuels, such as drying, extraction and fuel conversion. The method can be cost-effective, and several studies have demonstrated its potential. A recent study indicates that a biogas production level of 180.4 ml/g biogas, with a methane concentration of 65%, can be achieved using a two-stage anaerobic digestion process with different algae strains.

For macro-algae, anaerobic digestion is currently the most attractive conversion option. Depending on local circumstances, the biogas produced can be converted into bioelectricity or purified to replace natural gas. Biogas production can take place onshore or offshore, and can be optimized according to transport distance and availability of offshore gas, connection to the electricity grid from gas/oil platforms or wind turbines. Offshore gas production and shipment to land is also possible, but to reduce transport costs, pure methane needs to be upgraded and compressed offshore. Dedicated seaweed harvesting and processing vessels could be designed to optimize this process.

Biogas production is technically feasible for most macro-algae species, with advantages such as high productivity (e.g. *Ulva*: 100T sec/Ha/year, actual production), growth in seawater (no need for fresh water), and no competition with land-based agricultural surfaces.

9.7 Environment, pollution control :

Macro-algae help to reduce nutrient loads by naturally absorbing nitrates, phosphates, etc. This is of significant environmental interest for estuaries and eutrophic coastal zones. This is of significant environmental interest for estuaries and eutrophic coastal zones. The production of these macro-algae is seen as a means of reducing the impact of eutrophication caused by human activities (agriculture, urbanization, etc.) on environmental quality and biodiversity. It is also possible to treat effluent from human activities using macro-algae lagoons.

The wall polysaccharides (alginates, carrageenans, ulvans) present in algae have the ability to bind to heavy cations, enabling them to bind heavy metals; this gives them depolluting and detoxifying properties.

9.8 Conclusion :

Large seaweed production is based mainly on harvesting in natural environments and collection from strandings. Aquaculture predominates worldwide, particularly in Asia, where it has been mastered to perfection. This production method has strong development potential in France, particularly in Brittany, where a number of companies are beginning to invest in this field.

Compared to other sources such as plants, animals or bacteria, large algae and their polysaccharides offer great structural diversity, including a variety of branching and functional groupings, which give rise to numerous biological activities. This diversity, and the associated application potential, remains largely under-exploited from an industrial point of view.

With the ever-increasing consumption of prepared foods, interest in the properties of seaweed has intensified, particularly in the creation of a wide range of food texturizers designed to appeal to consumers. Today, the main aim of manufacturers is to appeal to consumers with innovative products. Phycocolloids (alginates, carrageenans, agars) have played a crucial role in changing eating habits. However, there are still many avenues to explore, not only in the food sector, but also in other sectors.

Indeed, as presented throughout this section, the applications of macro-algae have extended in recent years to the fields of cosmetology, pharmacology and, more generally, green chemistry and biotechnology, which are rapidly expanding markets.

10. Regulations :

10.1 Feeding :

10.1.1 French history :

In the late 1970s, the consumption of seaweed as a foodstuff in France began to grow, mainly under the impetus of vegetarian and macrobiotic consumers, as well as the arrival of migrant populations from Asia. At the time, seaweed was mainly imported from Asia. However, a number of entrepreneurs in Brittany are exploiting the wealth of seaweed on the Breton coast to offer products processed from local seaweed.

As early as the 1980s, companies began to organize themselves to harvest and process seaweed for food purposes. At the time, seaweed was considered a non-traditional foodstuff under current national regulations, and required prior authorization, granted following an opinion from France's Higher Council for Public Health (CSHPF). Several authorizations were issued, accompanied by opinions issued by the CSHPF, to regulate consumption and establish criteria for contaminants. At the time, 14 algae authorized for human consumption were included in a positive list.

In 1997, an important date in food regulation, the European regulation EC n°258/97 of January 27, 1997, known as "Novel Food", came into force. This regulation concerns novel foods and novel food ingredients. It applies in particular to seaweed for which it is not possible to prove traditional consumption in a member state of the European Union before the regulation came into force. In such cases, it becomes necessary to submit an application for authorization for use as a food or food ingredient, in accordance with the provisions of the above-mentioned EC regulation.

10.1.2 Algae currently used in France :

Table 11: Macro-algae authorized for human consumption.

Scientific name	**Common name**
Brown algae	-
Ascophyllum nodosum	-
Fucus vesiculosus	-
Himanthalia elongata	Sea spaghetti, sea beans
Undaria pinnatifida	Wakamé
Laminaria digitata	Kombu
Laminaria saccharina	Royal Kombu
Laminaria japonica	Kombu

Red algae	-
Palmaria palmata	Dulse
Porphyra umbilicalis	Nori
Porphyra tenera	Nori
Porphyra yezoensis	Nori
Porphyra dioica	Nori
Porphyra purpurea	Nori
Porphyra laciniata	Nori
Porphyra leucostica	Nori
Chondrus crispus	Pioca, lichen
Gracilaria verrucosa	Ogonori
Lithothamnium calcareum	Mäerl
Green algae	-
Ulva sp.	Sea Lettuce
Enteromorpha sp.	Aonori

Until 2010, it was possible to submit simplified dossiers for foods considered "substantially equivalent" to those already consumed or authorized. These dossiers were submitted to AFSSA (Agence française de sécurité sanitaire des aliments), which became ANSES (Agence nationale de sécurité sanitaire de l'alimentation, de l'environnement et du travail) in July 2010. It was also proven that the main imported Asian seaweeds had been consumed in Europe before 1997, which means that they can be used for human consumption in France, subject to compliance with regulatory criteria concerning contaminants such as iodine, heavy metals and microbial load.

There are currently 20 macro-algae authorized for human consumption in France. However, this list is not closed. It is still possible to obtain a simplified authorization to market seaweed with a recognized history of consumption in France or in other European Union countries, on condition that precise information is provided and following the opinion of the Direction Générale de la Concurrence, de la Consommation et de la Répression des Fraudes (DGCCRF). For non-traditional seaweed in the European Union, it is necessary to follow the procedure of EC regulation n°258/97 "Novel Food".

10.1.3 Restrictions on the use of algae for human consumption :

Alongside the list of authorized algae, the assessment bodies (previously CSHPF, then AFSSA, and now ANSES) have issued recommendations concerning the heavy metal and iodine content of food-grade algae. These algae must meet the criteria defined in the reference tables.

10.1.4 Maximum iodine content of edible seaweed :

Initially, a maximum threshold of 5,000 mg/kg was set for all algae, with specific consumption restrictions for kelp. In 2002, AFSSA issued an opinion lowering this threshold to 2,000 mg/kg for all algae, with the exception of kelp and *Gracilaria*, for which the threshold remained at 5,000 mg/kg.

The following consumption restrictions have been maintained: for adults, intake must be less than or equal to 30 mg of dry product per day, and for children under four, intake must be less than or equal to 15 mg of dry product per day (or 105 mg of fresh product).

Table 12: Maximum levels of heavy metals and iodine authorized in France for seaweed intended for human consumption.

Contaminant	**Algae vegetables or condiments (In mg/kg dry weight)**	**Food supplements (in mg/kg of product sold as is)**
Inorganic arsenic	3	-
Cadmium	0,5	3
Mercure	0,1	0,1
Lead	5	3
Pewter	5	-

Iodine	2 000	-

The table below summarizes the latest opinions and regulatory texts up to November 2010. In 2009, AFSSA published a new opinion on the maximum inorganic arsenic content recommended for kelp, and on how to consume kelp in view of its high iodine content. The maximum inorganic arsenic content of 3 mg/kg dry weight, recommended by the CSHPF on October 14, 1997, has been maintained. However, in its latest opinion, AFSSA recommended that the maximum threshold of 2,000 mg/kg iodine be applied to all species of edible seaweed. The exception for algae naturally rich in iodine (kelp, *Gracilaria*) has not been maintained. However, technological transformation processes can reduce the initial iodine content of seaweed, enabling kelp to be marketed in compliance with regulations. Although AFSSA's opinions are not regulatory texts, it is important to note that the DGCCRF can refer to them during controls, AFSSA's opinions then serving as a reference. There is no Community opinion on maximum iodine levels in seaweed.

In Germany, where several batches of Chinese seaweed have been withdrawn from the market, legislation sets a limit of 20 mg iodine per kg fresh weight for food products.

10.1.5 Maximum cadmium and lead content of algae :

EC regulation n°1881/2006 of December 19, 2006 set maximum levels for certain contaminants in foodstuffs, such as nitrates, aflatoxins, heavy metals, dioxins and polycyclic aromatic hydrocarbons. EC Regulation n°629/2008, amending EC Regulation n°1881/2006, specifically mentions seaweed for its natural ability to fix cadmium. This latest EU regulation sets maximum cadmium levels for food supplements made exclusively or mainly from seaweed. The maximum level is set at 3 mg/kg gross weight of the food supplement (container + contents), i.e. a higher value than that previously set by the CSHPF for seaweed (0.5 mg/kg dry).

EC regulation no. 629/2008 also sets a maximum lead content for food supplements in general, at 3 mg/kg gross weight. Unlike cadmium, this maximum limit is lower than the maximum limit set by the CSHPF for food algae (5 mg/kg dry weight). However,

for dietary supplements, the container must be taken into account, and the maximum content is given per kg gross weight of the dietary supplement.
Restrictions on the use of edible seaweed are therefore complex, as there are different maximum heavy metal values for fresh or dried seaweed consumed as a vegetable or condiment, and for seaweed used as a dietary supplement. The maximum heavy metal values for food supplements are European values, while the maximum values for fresh or dried seaweed consumed as vegetables or condiments are levels set at national level in France.

10.1.6 Microbiology :

To conclude these regulatory aspects, bagged seaweed, in the form of dry products, must comply with the microbiological criteria set out in the CSHPF opinion. It is also important to remember the EU provisions in force concerning microbiological criteria for foodstuffs. These standards are designed to ensure that products do not contain dangerous levels of pathogenic micro-organisms, thereby guaranteeing consumer safety.

Table 13: CSHPF microbiological criteria.

Microbiological criteria	**Maximum permitted values**
Mesophilic aerobic germs	$\leq 10^5$ / g
Faecal coliforms	≤ 10 / g
Sulfiting anaerobes	$\leq 10^2$ / g
Staphylococcus aureus	$\leq 10^2$ / g
Clostridium perfringens	≤ 1 / g
Salmonella	Absence in 25 g

10.1.7 Food-grade algal colloids :

Polysaccharides derived from macro-algae, previously presented as food colloids, are listed under European food additive codes.

Table 14: European food additive codes for algal colloids.

Colloid name	**European food additive code**
Alginic acid	E 400
Sodium alginate	E 401
Potassium alginate	E 402
Ammonium alginate	E 403
Calcium alginate	E 404
Propylene glycol alginate	E 405
Agar agar	E 406
Carrageenan	E 407
Semi-refined carrageenan	E 407 a

With regard to carrageenans, it is important to mention the toxicological data associated with them. The Acceptable Daily Intake (ADI) is set at 75 mg/kg. At commonly used doses, no adverse effects have been observed. However, some studies suggest that these compounds may act as immunosuppressants, potentially capable of disrupting antibody production and anti-tumor immunity. In addition, degraded carrageenans are known to have an ulcerogenic effect in the colon, even at moderate doses. Consequently, the use of carrageenans that have undergone chemical modifications leading to partial degradation is strictly prohibited.

10.2 In cosmetics :

The regulations applicable to macro-algae in the cosmetics sector are identical to those for micro-algae. To date, there are no specific regulations for "algae" in cosmetics. The valorization of macro-algae must comply with general regulatory constraints and

classic official recommendations, such as claims, certifications, REACH recommendations, and nomenclatures for cosmetics. In order to bring about regulatory change in this sector, the major groups have joined forces at European level to propose good manufacturing practices with the ISO 22716 standard, and to support the introduction of stricter regulations for cosmetic products by 2013, in particular with the REACH regulation (EC no. 1223/2009, which came into force on July 11, 2013).
ISO/TC 217 and ISO/NP 16128 define technical criteria for "natural" and "organic" cosmetic ingredients and products. For cosmetic extracts derived from macro-algae, a description of the extract used is sufficient; it is not mandatory to specify the species of algae used.
The INCI nomenclature, initiated in 1973 by the C.T.F.A. (Cosmetic, Toiletry and Fragrance Association) and made mandatory in Europe since 1998, informs consumers of the composition of ingredients in descending order of concentration, under an internationally recognized name.

10.3 Other sectors :

There are also specific regulations governing the use of algae as fertilizers (NF U 42-001), organic soil improvers (NF U 44-051), calcomagnesium soil improvers (NF U 41-001; NF U 44-203), and culture media (NF U 44-551; NF U 44-571).

10.4 The organic "seaweed" sector :

Since the entry into force of EC regulation no. 710/2009, seaweed production has been governed by the European Organic Farming Regulation. Currently, around ten operators in Brittany's seaweed sector already hold organic certification, whether for foodstuffs containing seaweed, inputs (such as soil conditioners or "phytostimulants"), or cosmetics (under private specifications).
Organic food products based on seaweed are varied: fresh salted seaweed, seaweed salt, seaweed tartar, seaweed bread and cheese, dehydrated seaweed flakes, etc. A Breton company also markets organic agar-agar as a gelling agent. As for companies

incorporating seaweed into cosmetics, they are certified to private specifications such as those of Cosmébio, Nature & Progrès, or BDIH.
The European Organic Regulation, interpreted by the INAO (French reading guide), stipulates that seaweed can only be certified organic if it is harvested or cultivated in bodies of water classified as "good ecological status" or "very good ecological status" according to the Water Framework Directive (WFD - 2000/60), and if it also meets other sanitary and chemical criteria.

CHAPTER 5: ANALYSIS & OUTLOOK

1. Restitution of the Thematic Reflection Workshops :

The main points to emerge from the thematic workshops (ART) held at the "Algae: sectors of the future" symposium are summarized below.

1.1 ART 1: Functional genomics programs on algae: an asset for biotechnologies :

1.1.1 Current situation :

On a global scale, algae research, both macro and micro, benefits from a solid base of diversified skills. Leading research centers such as CNRS, CEA and Ifremer in France, as well as similar institutions elsewhere in the world, are home to numerous research teams. These institutions publish extensively in leading journals, even though they face intense international competition. The USA and Asia, in particular, are major players in this field, with considerable investment in algae research.

We are at a crucial time when the potential of algal research needs to be better structured and strengthened to meet the major challenges of genomics, particularly in terms of industrial property. Better coordination and greater visibility of international initiatives are needed to make progress in this field.

In the context of modern biology, access to the genomes of organisms of interest is essential for developing functional genomics programs. Mass sequencing programs for algal strains are currently underway worldwide, and need to be continued. The strategic challenge in terms of industrial property is considerable, as current knowledge is still fragmentary. At present, the species exploited are often harvested directly from their natural environment or cultured from strains taken from the wild. As in agriculture, where cultivated varieties have characteristics far removed from those of their wild ancestors, it will be necessary to implement a domestication and selection process to develop algae varieties optimized for specific applications. This process will need to draw on advances in genomics if it is to be effective and rapid. The following proposals have been put forward:

1.1.2 Proposals

1.1.2.1 Developing the Domestication of Species of Aquacultural and Biotechnological Interest

- ✓ **Adopt a Global Approach to Culture**: Favor an integrated approach that takes into account conflicts of use and integrated aquaculture.
- ✓ **Implement Varietal Selection**: Target not only the production of compounds of interest through metabolic reprogramming, but also adaptation to cropping systems, by controlling life cycles and other relevant factors.
- ✓ **Anticipating phytopathological problems**: Take steps to ensure the sustainability of production by anticipating phytopathological problems.

1.1.2.2 Supporting Knowledge Production on Model Organisms

- ✓ **Exploring Natural Biodiversity**: Increase efforts to explore and understand the natural biodiversity of algae.
- ✓ **Strengthen knowledge of key models**: Develop a solid knowledge base of the biology of a few key models.
- ✓ **Providing Scientific Bases for Varietal Improvement**: Establishing scientific foundations to rationalize and optimize varietal improvement programs.

1.1.2.3 Moving from Functional Genomics to Systems and Integrative Biology

- ✓ **Strengthen the training of young researchers**: Improve the training of young researchers in this innovative sector.
- ✓ **Developing multidisciplinarity**: Encouraging collaboration between different disciplines, such as biology, chemistry, ecology, mathematics and bioinformatics.
- ✓ **Establish a Sequencing Roadmap**: Develop a plan for the sequencing of new micro- and macro-algae species.
- ✓ **Develop Genetic Tools and "Omics" Approaches**: Create "omics" tools and methods for species of interest that are not yet accessible to these techniques.

1.2 ART 2: Seaweed cultivation and harvesting :

1.2.1 Current situation :

The Adebiotech conference revealed the disparities between the macroalgae and microalgae sectors. Macroalgae, used in large quantities (millions of tonnes) in markets such as agri-food, also have potential for the manufacture of biomaterials in the short term. Microalgae, on the other hand, are produced in much more limited quantities (10,000 tonnes per year). Their development is mainly held back by the high cost of production and processing, limiting their use to high value-added sectors such as hatcheries, cosmetics and nutraceuticals.

Although macroalgae are widely used in the food industry, they remain too expensive for energy applications. This raises questions about the technical and economic challenges facing these sectors.

1.2.2 Production optimization :

Photosynthetic algae production depends on the interaction between three essential elements:

- ✓ A biological organism
- ✓ A growing medium
- ✓ Light, whose intensity and quality vary over time

To optimize this production, it is crucial to improve the management of these interactions. This includes the development of bioprocesses that integrate all aspects of this interaction: optimization of strains and culture conditions for increased economic efficiency, light management (including diffusion within cultures and the use of wavelengths not required for photosynthesis, such as infrared), and biomass harvesting and biorefining to maximize value.

Fermenting cultures, which do not require light, may offer a solution to the challenges of light management. However, their limitations and markets remain to be explored in greater detail.

1.2.3 Separation and Recovery challenges :

Separating algal biomass from its culture medium, usually by centrifugation, remains a costly operation. Although other technologies have been tested over the last two decades, none has yet been widely adopted by microalgae producers.

In addition to technological advances, two other approaches can improve the technical-economic balance of algae projects:

- ✓ **Improved product valorization**: Finding better ways to valorize products derived from algae.
- ✓ **Use of Industrial Effluents**: Replace crop inputs (carbon, nitrogen, phosphorus) with industrial effluents, which could also generate additional income.

This workshop will focus mainly on the second approach, the use of industrial effluents.

1.3 ART 3: Industrial flow and process management Current status :

1.3.1 Current situation :

The main obstacles to the industrial production of microalgae lie in the cultivation and harvesting phases. Lifecycle analyses show that the energy consumed comes mainly from electrical sources, used for biomass agitation and separation, as well as for CO_2 injection. Among the major challenges is the separation of algal matter from interstitial and compositional water, a particularly energy-intensive operation. Although technologies exist, they are not always suitable for large-scale extrapolation, requiring a sufficient quantity of biomass to adjust current processes. Additional constraints include those linked to the use of seawater, where the presence of salts requires specific equipment and processes. The challenges of $CO_{(2)}$ injection and temperature control are also complex to overcome in an open environment, due to the influence of atmospheric elements such as sun, rain and wind. What's more, algae growth requires the supply of nitrogen, phosphorus and other mineral elements, in addition to water and CO_2. It is therefore crucial to analyze the path of these elements to maximize their recycling. Closure of flows is essential to reduce input-related production costs and lower the environmental impact of the process.

1.3.2 Proposals :

To improve microalgae transformation processes, initial life cycle analyses reveal that it is necessary to :

- ✓ Optimize growing and harvesting systems to reduce energy consumption.
- ✓ Study the recycling and recovery of all flows, such as water, mineral elements and CO_2.
- ✓ Prefer and examine cracking and refining of all recoverable materials, taking into account the high costs associated with algae production and processing, which requires the development of breakthrough innovations.
- ✓ Set up complete industrial biorefinery pilots, similar to those in the plant-based biofuels sector, to validate concepts, assess impacts and specify production costs.

1.4 ART 4: Adding value to seaweed in high value-added markets :

1.4.1 Current situation :

For many years now, algae have been the focus of developments in the cosmetics, nutraceuticals and food sectors, for both human and animal consumption. During this thematic workshop, several concrete examples illustrated these advances. One notable example was the presentation of the successful development of a micro-algae as a nutritional supplement, particularly from the angle of EU regulations. The use of microalgae as cell reactors for the production of molecules of medical interest, such as recombinant proteins (biological markers, vaccines, etc.), or of cosmetic and nutraceutical interest, was also highlighted. In the animal feed sector, and more specifically in aquaculture, whole microalgae have long been used as feed for rotifers, fry and fish. An industrial activity in this field was described, enabling a concrete appreciation of the conditions of production, the species cultivated and the economic stakes involved.

In addition, various metabolic pathways from primary and/or secondary metabolism have been described, opening up the possibility of obtaining molecules of interest in the fields of health and nutrition, for example in oncology and immunology. In France, the algae industry has to overcome a number of scientific, technological and regulatory hurdles if it is to develop fully. However, an essential step is the identification of market needs, which determines the commercial and economic success of an original natural molecule. This task is made all the more complex by the fact that the potential fields of application for micro- and macro-algae are diverse, due to the biodiversity of algae, which remains incompletely inventoried to date.

Once a need and a market have been identified, the next step is to remove the scientific barriers, in particular by identifying the species and strains likely to be of interest, in terms of both the production of specific active ingredients and the feasibility of cultivating them. Selecting a productive strain is of little use if it proves difficult to cultivate. Furthermore, it may be necessary to over-express certain biochemical synthesis pathways to produce metabolites of interest, a challenge in itself. On the technological front, it is crucial to implement production, extraction and purification steps that enable the desired material to be obtained economically and in compliance with current legislation (for example, complying with the REACH directive in the case of solvent extraction, taking residual traces into account). With this in mind, we need to consider the development of "green" extraction technologies, along the lines of those used in the traditional plant industry.

On the other hand, scientific, technological and economic challenges are accompanied by intellectual property constraints, an aspect often overlooked by researchers. Identifying a strain of interest or an efficient production process is of little interest if they are not free to exploit. If this is the case, it is imperative to study the possibility of protecting the invention by filing a patent application, provided that sufficient evidence of the inventive nature of the work carried out can be provided. These considerations require the intervention of experts, and represent time-consuming and costly services, particularly for small companies. The recent structuring of scientific organizations (CNRS, INRA, INSERM, etc.) and academic research establishments (Universities, Schools) in the field of research and technology transfer is a major asset in overcoming this constraint.

Finally, bringing a material or active ingredient to market requires compliance with current regulations. This aspect of development is often underestimated, both in terms of legal obligations and costs. The preparation of marketing authorization applications is generally long and costly. This thematic workshop therefore led to the formulation of the following proposals.

1.4.2 Proposals :

Finding 1: It is crucial to identify emerging markets as a matter of priority, as well as those of particular interest for innovative active ingredients derived from algal biomass.

- ✓ **Proposal 1:** Create a business intelligence organization dedicated to the algae industry, covering the national, European and international levels.

Finding 2: Although the creation of institutes and the development of technology platforms are underway, visibility for industry remains limited.

- ✓ **Proposal 2:** The French government should support research by encouraging scientists to collaborate and pool their resources within a given territory or region. Technology platforms should include all the stages essential to the industry, such as extraction, purification, characterization and, potentially, activity screening. Academic initiatives aimed at creating and developing extract libraries or marine extract libraries accessible to industry, as well as setting up "phycobanks" dedicated to microalgae, should also be supported.

Finding 3: Do consumers see the link between raw biomass and the assets it contains?

- ✓ **Proposal 3:** Implement an asset traceability system throughout the process to ensure the security of all players in the industrial chain and boost consumer confidence in the quality of seaweed-derived products.
- ✓ **Additional proposals :**
- ✓ **Support for intellectual property:** The French government should reinforce the creation of a solid intellectual property base in France by equipping development institutes, R&D platforms and industrial pilots with structures capable of helping companies in the micro- and macro-algae sector to develop and protect their technology. These structures should be able to carry out detailed bibliographies, freedom-to-operate analyses, prior art searches, patent mapping and patent drafting.
- ✓ **Support for marketing initiatives:** The French government should also support innovative initiatives to bring micro- and macro-algae-based materials and active ingredients to market, by assisting companies, particularly SMEs and VSEs that are very active in this field. This includes :
    - Funding for preliminary regulatory studies.
    - Support for the preparation of regulatory approval files, potentially via shared structures (platforms, institutes) in perfect synergy with the sector's competitive clusters.
    - Support for the industry's collective efforts with European regulatory authorities, for example, by backing the request to create an organic label for macro-algae intended for food use.

1.5 ART 5: Power generation and environmental applications :

1.5.1 Current situation :

Against a backdrop of widespread media coverage, although the concept of algofuels appears to be a promising energy option for the future, it is clear that a great deal of research and development will be required to make the algal sector economically viable on an industrial scale and to improve energy balances. The scientific community estimates that it will take 10 to 15 years for third-generation biofuels to reach technical maturity.

From a research and development point of view, the expansion of the algal concept could depend on a thorough exploration of microalgae biodiversity, a significant improvement in the productivity and stability of cultivation processes, and optimization of harvesting and extraction systems, all in parallel with a reduction in the costs and energy required. It will be essential to go through a pre-industrial phase before aiming for large-scale production, focusing initially on the production of high-value-added molecules. This approach will need to integrate the biorefinery concept and examine the opportunities offered by the use of liquid or gaseous inputs from industrial recycling, such as CO_2, nitrogen and phosphorus.

A group of manufacturers involved in CO_2 production, the water and marine industries, as well as the aeronautical, land and sea transport, and petrochemical sectors, working together under the aegis and with the support of the French government, will certainly be necessary to enable the emergence of an integrated industry.

Important levers and success factors have been identified. Political players will have to be heavily involved and play an essential role in the implementation of the value chains, with regard to conflicts of use, land-use planning and the regulation of the exploitation of the national maritime and continental aquatic heritage for the production of algal biomass. The social acceptability of products from this sector will also have to be examined with particular attention, especially in the context of biorefineries.

Last but not least, the choice of processes will have to take into account not only economic aspects, but also environmental impacts, in particular greenhouse gas emissions and the effects on resources and ecosystems.

1.5.2 Proposals :

A global approach that takes all the above aspects into account will lead to the development of new, evolving business models, in line with technological advances (yields, standardization stages) and the new markets to be integrated and developed. To tackle these challenges successfully and raise the profile of this activity on an international scale, a number of proposals were put forward at Workshop 5, "Energy production and environmental applications":

- ✓ Establish a national roadmap to guide and structure efforts in the algofuels and environmental applications sector.
- ✓ Create a common structure for coordinating, coordinating and leveraging public and industrial research and development (R&D) efforts, to maximize synergies and project efficiency.

- ✓ Set up large-scale demonstrators in parallel with fundamental research, ensuring dynamic interaction between the two to accelerate the application of innovations.
- ✓ Set up an appropriate training program to compensate for the lack of technicians, engineers and researchers in this multi-disciplinary field, essential to the development of the sector.
- ✓ Anticipate and activate the legislative and regulatory levers needed to support and frame the development of technologies linked to energy production from algae and environmental applications.

1.6 ART 6: Chemistry and agro-materials :

1.6.1 Current situation :

Discussions at the Chemistry and Agro-Materials ART led to the following conclusions:

Microalgae and macroalgae represent a biomass offering an interesting diversity of compounds, with a notable complementarity and significant originality compared to compounds derived from terrestrial biomasses. Industries serving the chemical and agro-materials markets require a constant supply in terms of quantity and quality.

For macro-algae, the extraction industry is already well established, notably with the production of food and technical texturizers, as well as extracts used in agriculture, animal nutrition and cosmetics. However, it will be necessary to scale up to meet the growing needs of the chemical and materials industries. This challenge is less pronounced for macroalgae than for microalgae.

Biomass can only be sourced from crops whose cultivation methods are well understood and controlled, so as to guarantee a sufficient quantity of renewable material of consistent quality. The naturally high productivity of algae should make it possible to obtain significant biomass without the use of genetically modified organisms, which will help to promote societal acceptance of their use.

The use of seaweed would also considerably reduce the consumption of freshwater resources, as seawater is abundantly available in coastal areas or in offshore systems.

In Western countries, it will probably be easier to develop chemistry from algae than to get consumers used to incorporating them into their diet. This approach would make it possible to reserve agricultural land for food crops, while using the seas to provide intermediates for chemicals and materials.

Lastly, there is already a solid base of knowledge at national level on algal matter and the chemistry of its constituents, which should be favorable to the development of this sector in the medium term.

1.6.2 Proposals :

- ✓ Develop high-quality, low-cost algal production in sufficient quantities to meet industrial needs.
- ✓ Continue to explore the agrochemical potential of algal biomass to discover new applications and opportunities.
- ✓ Identify the main value-added sectors that will make the sector profitable and ensure its long-term economic viability.

2. Perspectives

Around the world, many regions benefit from climatic conditions favorable to seaweed cultivation, as well as vast maritime coastlines conducive to this activity. With thousands of kilometers of coastline spread across the globe, many countries have extensive Exclusive Economic Zones, offering rich biodiversity in their coastal ecosystems (salt marshes, lagoons, shallow bays, etc.), ranging from tropical to temperate to polar climates. This diversity is matched by the presence of numerous species of marine plants with potential for cultivation and development.

On a global scale, several nations stand out for their leadership in scientific publications and patent filings in the field of algae science and technology. Research and development (R&D) skills are varied and cover the entire value chain, from genomics to biorefinery, including process engineering for cultivation, harvesting and processing. Several academic and industrial players have established themselves as international benchmarks, notably through major collaborative projects supported by competitiveness clusters and co-financed by various public and private agencies.

The algae industry is already well established in several regions of the world, with companies operating in the macro-algae markets for food, agriculture and cosmetics, as well as in the micro-algae markets for nutraceuticals and aquaculture. These companies, often small or medium-sized, are heavily involved in innovation and play a crucial role in numerous R&D projects. This dynamic shows a real willingness on the part of companies to develop their activities in this sector by adapting their skills or creating new activities.

To structure this sector on a global scale, a number of international initiatives and partnerships have been set up, bringing together various academic, industrial and institutional players. These collaborations aim to support structuring projects for the sector, whose ambitions are to reach industrial scale.

In addition, clusters and groups of players have been set up in various parts of the world to federate research and innovation efforts. These initiatives testify to the international enthusiasm for the development of the algae industry, and the interest of companies from all sectors in participating in its growth.

However, the whole world suffers from the dispersion of its academic skills and the weakness of its technological facilities, often divided between numerous large research organizations. Private-sector structures, often small-scale, do not always have the resources to go beyond their technical and financial possibilities. What's more, several regions lack industrial demonstrators of significant size, making them less competitive with major powers such as the USA and certain Asian countries. This lag is mainly due to a lack of public and private investment, which is hampering the overall development of the sector.

The wait-and-see attitude of major industrial groups in many regions is also holding back the industrialization of processes, despite the fact that a global industrial sector cannot be built without their commitment. Although several countries are leaders in terms of publications and patents, they lag far behind the USA, China and even the UK, where over two billion dollars have already been invested in this sector.

The USA, in particular, has attracted 80% of global investment in algae, with industrial giants such as Exxon Mobil, Shell, and HR Petroleum involved in large-scale projects. These investments are essential if we are to overcome the technical and economic challenges identified. On a global scale, public-private co-investment could be a key to success for the algae industry, accompanied by a global Intellectual Property strategy. Indeed, the most aggressive countries in this field, notably the United States,

China and Japan, are locking in technology through a large number of patents, making the industry profitable for market leaders.
For algae to become a viable raw material for mass markets (chemicals, environment, energy), it is necessary to reduce production and processing costs, and to remove the technological barriers that limit large-scale production. Problems linked to the use of fossil fuels and the preservation of the environment are driving the exploration of new alternatives in terms of energy production and consumption on a global scale.
In this context, algae, both macro and micro, can represent an interesting alternative for various sectors (agri-food, cosmetics, environment, bioenergy).
With the support of research organizations and stakeholder groups, global initiatives must structure research on an international scale.
This will make it possible to leverage academic skills to achieve realistic, shared industrial objectives, and to prepare the laws and regulations needed to integrate this sector into society and the regions concerned. Reinforcing and setting up technical centers, technology platforms and centers of excellence will help federate players and give research the means to meet the demands of industry. Scaling up and transferring technology from the laboratory to industry will only be possible if everyone is involved in collaborative R&D programs and industrial demonstrators. These tools will make it possible to test, optimize and validate technologies developed on a small scale, before moving on to economically viable industrial installations.
Finally, this structuring of R&D forces will encourage major groups to invest upstream in projects, and create a genuine link between research and industry, catalyzing the development of the algae sector on a global scale.

3. Global players in the algae industry :

From this preliminary work, it appears that the players - both research laboratories and companies specializing in micro- and macro-algae - are mainly concentrated in certain coastal regions, such as the Mediterranean rim, East Asia, North America and Northern Europe. Macro-algae companies are mainly located in regions such as Brittany in France, Norway and Japan, where climatic and environmental conditions favor their cultivation. Micro-algae companies, notably spirulina producers, are present in more diversified regions, including Mediterranean areas, South-East Asia, North America, and even continental regions far from sea coasts, such as parts of the USA, Brazil and Australia.

The studies carried out, based on a partial census of the sector, show the distribution of players according to their type of activity, in particular research and companies specializing in macro- and micro-algae, as well as the distribution of research laboratories along the value chain. Half of the players surveyed are companies operating in the microalgae sector.

It is also important to note that many small companies in the macro-algae sector have not been included, due to their small size and the difficulty of identifying them. Around a third of the world's laboratories involved in algae value chains focus on the study of strains, covering various scientific fields such as genomics and metabolomics, while other research activities are relatively well distributed along the value chain.

4. Ongoing programs and cooperation :

The appendix lists the various programs currently underway, accredited by competitiveness clusters and international consortia, along with the partners involved and the budgets allocated. These projects illustrate the close cooperation between industry and public-sector laboratories worldwide. A total of 51 projects were listed, with an overall budget of at least 350 million euros. A quick analysis of these data reveals that the vast majority of project budgets are devoted to micro-algae (88%), with a higher average budget than that allocated to macro-algae. A similar breakdown is observed when comparing the distribution between "basic research" and "development", with two-thirds of development projects accounting for 90% of funding. The average budget is therefore higher for development projects than for basic research projects.

5. Analysis of the global algae industry :

A detailed analysis of the two algae sectors, microalgae and macroalgae, is presented in the following pages. For each sector, the value chain is described, with a typology of the players involved and the technological obstacles identified at each stage. This is followed by a presentation of current and potential applications for algae in high value-added and mass markets, along with the advantages and challenges specific to each application sector. Finally, this section includes a SWOT (strengths, weaknesses, opportunities, threats) analysis for each of the two sectors.

The micro-algae sector, which is less developed than the macro-algae sector, is the subject of a study to identify the markets accessible in the short, medium and long term, depending on the technological barriers that have been overcome.

Strategic proposals: Support the emergence and competitiveness of the algae industry worldwide.

6. Take stock of the algae sector and identify technical and economic challenges

- ✓ **Objectives :**
    - Identify the different ways of using algae and the associated constraints (cost, quality, product traceability, environmental impact, energy balance, etc.).
    - Harmonize the costs of products from the supply chains with the target markets.
- ✓ **Actions :**
    - Draw up a precise inventory of existing and emerging sectors in terms of players (research organizations, major groups, SME-SMIs), regulations, industrial property, concerted actions already undertaken, and structuring projects.
    - Conduct specific studies for each sector: market studies and technical-economic feasibility studies (sector maturity, implementation deadlines, needs).

- Develop specific roadmaps for each sector: processes to be developed, R&D efforts, legislative framework, planning, current and future markets to 2020.
- Consider setting up a product/co-product exchange to form a chain similar to the biorefinery concept, where a co-product from one sector could be used as a raw material by another.

7. Accelerate the innovation process to remove technological barriers :

Five priority work areas have been identified:

7.1 Selecting species and strains adapted to the needs of the industry :

- ✓ **Objective:** Identify and optimize strains suitable for industrial use.
- ✓ **Actions :**
  - Identify public and private stock libraries worldwide.
  - Structuring the network of strain libraries and the tools needed to use them and make resources more reliable (accessibility, traceability).
  - Create a directory of laboratories working or likely to work on active ingredient screening for the food, cosmetics and pharmaceuticals sectors.
  - Encourage R&D projects focusing on knowledge of species and their biodiversity.
  - Develop strain improvement technologies, sequence genomes and develop key models.

7.2 Understanding the accumulation mechanisms of compounds of interest :

- ✓ **Objective:** Understand the link between environmental conditions, intracellular accumulation and overall productivity.
- ✓ **Actions :**
  - Research and study conditions favoring the accumulation of compounds of interest.
  - Identify the activity of key genes during accumulation phases.
  - Assessing and optimizing the effects of deficiencies in the natural environment.
  - Develop an integrative biology approach to studying organisms as a whole.

7.3 Improve cultivation, harvesting and extraction processes by optimizing their productivity and limiting their impact:

- ✓ **Objective:** Increase productivity while minimizing economic and energy costs, and environmental impact.
- ✓ **Actions :**
  - Conduct an industrial watch on available and exploitable processes for the algae industry.
  - Develop production systems adapted to different sectors, from small-scale production to large-scale applications.
  - Optimize intensive and extensive cultivation by controlling climatic factors and automating processes.
  - Develop sustainable biomass harvesting and processing techniques, as well as flow management processes.

7.4 Process scaling up, techno-economic analysis and impact assessment :

- ✓ **Objective:** more accurately predict achievable productivity rates, associated impacts and costs, and define business models.
- ✓ **Actions :**
  - Study and assess the availability and toxicity of industrial effluents.
  - Implement integrated biological control approaches to limit contamination.
  - Develop crop monitoring and risk anticipation tools.
  - Assess the risks and impacts on natural ecosystems associated with large-scale cultivation.
  - Collect and validate data for Life Cycle Assessment of production systems.

7.5 Developing application synergies: industrial ecology and the circular economy :

- ✓ **Objective:** capitalize on the environmental and waste treatment potential of algae, and bring them to the right industrial markets.
- ✓ **Actions :**
  - Integrating algae cultivation into industrial ecosystems for effluent treatment and CO_2 remediation.
  - Integrate algae as a raw material in biorefinery processes developed for other agro-resources.

8. Encourage the development of equipment manufacturers :

- ✓ **Objective:** Develop industrial capacity to build seaweed production and processing facilities that meet the needs of different sectors and markets.
- ✓ **Actions :**

- Develop the design, manufacture and installation of photobioreactors, raceways, open ponds and large-scale fermenters.
- Set up demonstrators to remove the technological barriers to large-scale cultivation.
- Develop harvesting tools and pre-treatment facilities, as well as biorefinery plants.
- Involve equipment manufacturers from other sectors in the projects, so that they can acquire the skills needed to build large-scale installations.
- Encourage patenting of developed technologies.

9. Work on the integration of companies into the environmental and social landscape:

Three priorities have been identified for the effective integration of companies in the environmental and social context.

9.1 Identify regulatory constraints, resource requirements, and characterize territorial barriers :

- ✓ **Objectives:** Identify potential locations for new businesses and understand the constraints associated with these sites.
- ✓ **Actions :**
  - Support regulatory procedures (water laws, ICPE, etc.).
  - Define the ideal geographical conditions for the site: proximity to seawater, freshwater, access to land, sunshine, temperature, and synergies with industrial, agricultural and aquaculture activities.
  - Identify the upstream technological environment, such as the proximity of laboratories, companies and effluents.

- Identify the downstream logistics environment (product end-users) to minimize carbon footprint and environmental impact.
- Identify and locate production sites subject to regulations.
- Extend these criteria to the French overseas departments and territories and worldwide (lagoons, hypersaline zones, etc.).

9.2 Assessing and mitigating business impacts :

✓ **Objective:** Implement an industrial strategy for sustainable development.

✓ **Actions :**

- Study environmental acceptability (potential impacts on natural environments) and societal acceptability (management of conflicts of use); develop methods to minimize environmental impact.
- Harmonize Life Cycle Assessment (LCA) techniques and databases to facilitate the establishment of an LCA for each process and the industrial value chains envisaged.

9.3. Make training organizations aware of industry's need for engineers and technicians :

✓ **Objective:** Improve the range of training courses on offer and ensure that graduates' skills match the needs of the industry.

✓ **Actions :**

- Anticipate business demand for technicians, engineers and PhD students.
- Align training courses with market demand and needs.
- Develop continuing education programs and specific modules for companies.
- Promote exchanges between research/training establishments and industry.

10. Support the structuring of the algae industry :

Three priorities have been identified to reinforce the structuring of the algae industry.

10.1 Creating shared R&D tools :

- ✓ **Objective:**
  - Synergize the skills of different players (academic, industrial).
  - Enable project leaders to transfer their technologies to pre-industrial scale.
- ✓ **Actions :**
  - Promote the creation of shared tools, platforms and demonstrators covering all stages of the value chain.
  - Strengthen banks of algal strains and extracts.
  - Make coherent use of calls for collaborative projects from the French government, the regions and Europe, with the support of competitiveness clusters.

10.2 Promote the sharing of monitoring and benchmarking tools :

- ✓ **Objective:** Strengthen the use of local innovation support services by companies and laboratories.
- ✓ **Actions :**
  - Create a watchdog unit to monitor patents, publications and international players (e.g.: benchmark study on global clusters working on micro-algae, launched by the relevant competitiveness clusters).
  - Reinforce the monitoring and dissemination of information on calls for projects in collaboration with the competitiveness clusters.
  - Develop an Industrial Property support strategy.
  - Help with regulatory and/or toxicological procedures, particularly in the food sector.
  - Mobilize existing innovation systems to support the algae industry.

10.3 Identify or create the appropriate structure to represent the algae industry at the European Algae Biomass Association (EABA)

- ✓ **Objective:** To meet the need for a meeting place to structure, develop and animate the industry.
- ✓ **Actions :**
  - Identify existing structures in the macro- and micro-algae sectors and establish links with EABA to coordinate efforts on a European scale (collaborations, responses to European calls for projects).

GENERAL CONCLUSION

Algae, often perceived as simple aquatic organisms, reveal an unsuspected complexity and potential through an in-depth exploration of their characteristics, applications and impact on various sectors. This study highlights their crucial role in marine ecosystems, as well as their capacity to offer innovative solutions to contemporary environmental and economic challenges.

In the course of this research, it became clear that algae, whether micro or macro, are much more than a source of oxygen or a component of the food chain. They are proving to be versatile resources, capable of meeting a multitude of human needs, from renewable energy production to food, cosmetics and medicine. Their potential in the fight against climate change, thanks to their ability to capture carbon dioxide and produce biomolecules, is particularly promising.

Innovation in algae cultivation and exploitation techniques has opened up new prospects, making large-scale production not only possible, but also economically viable. However, implementing these innovations on an industrial scale requires close collaboration between researchers, industrialists and political decision-makers, as well as an appropriate regulatory framework to guarantee the sustainability of marine ecosystems.

In conclusion, this study underlines the importance of algae as a major player in sustainable development. They represent a potential response to current global challenges, particularly in terms of food security, energy transition and environmental protection. To fully exploit this potential, it is essential to pursue research and innovation efforts, while integrating algae into global economic and environmental strategies. The future of algae will depend on our ability to recognize their value and integrate them into our production systems in an environmentally-friendly and economically viable way.

BIBLIOGRAPHICAL REFERENCES

1. Ainane, T., & Abourriche, A. (2014). Brown seaweed *Bifurcariabifurcata*: Bioguided fractionation of extracts by antibacterial activity and cytotoxicity test. *Biosciences Biotechnology Research Asia, 11*, 1081-1085.
2. Ainane, T., Abdoul-Latif, F., El Yaagoubi, B., Boujaber, N., Oumaskour, K., &Ainane, A. (2021). *Nannochloropsisoculata* microalga for fertilization: Bioguided fractionation of N-hexane extract by stimulating growth activity for *Cucurbita moschata*. *PharmacologyOnline, 3*, 1803-1809.
3. Ainane, T., Abourriche, A., & Kabbaj, M. (2015). Physico-chemical analysis by SEM-EDX and FTIR two brown algae *Cystoseiratamariscifolia* and *Bifurcariabifurcata*. *BioTechnology An Iridian Journal, 11*(5), 185-188.
4. Ainane, T., Abourriche, A., Bennamara, A., Talbi, M., &Lemrani, M. (2015). Anti-leishmanial activity of extracts from a brown seaweed *Bifurcariabifurcata* from the Atlantic coast of Morocco. *Phytothérapie*, 1-6.
5. Ainane, T., Abourriche, A., Kabbaj, M., Elkouali, M., Bennamara, A., Charrouf, M., &Talbi, M. (2014). Removal of hexavalent chromium from aqueous solution by raw and chemically modified seaweed Bifurcariabifurcata. *Journal of Materials and Environmental Science, 5*(4), 975-982.
6. Ainane, T., Abourriche, A., Kabbaj, M., Elkouali, M., Bennamara, A., Charrouf, M., &Lemrani, M. (2014). Biological activities of extracts from seaweed *Cystoseiratamariscifolia*: Antibacterial activity, antileishmanial activity and cytotoxicity. *Journal of Chemical and Pharmaceutical Research, 6*, 607-611.
7. Alam, M. A., Xu, J. L., & Wang, Z. (Eds.). (2020). *Microalgae biotechnology for food, health and high value products* (pp. 81-123). Springer.

8. Alam, M. A., Xu, J. L., & Wang, Z. (Eds.). (2020). *Microalgae biotechnology for food, health and high value products* (pp. 81-123). Springer.
9. Babich, O., Sukhikh, S., Larina, V., Kalashnikova, O., Kashirskikh, E., Prosekov, A., ... &Dolganyuk, V. (2022). Algae: Study of edible and biologically active fractions, their properties and applications. *Plants, 11*(6), 780.
10. Barsanti, L., & Gualtieri, P. (2022). *Algae: Anatomy, biochemistry, and biotechnology*. CRC Press.
11. Fernández, F. G. A., Reis, A., Wijffels, R. H., Barbosa, M., Verdelho, V., & Llamas, B. (2021). The role of microalgae in the bioeconomy. *New Biotechnology, 61*, 99-107.
12. Fernández, L. A. G., Ramos, V. C., Polo, M. S., & Castillo, N. A. M. (2023). Fundamentals in applications of algae biomass: A review. *Journal of Environmental Management, 338*, 117830.
13. Harvey, W. H. (2024). *A manual of the British algae*. BoD-Books on Demand.
14. Jacob-Lopes, E., Queiroz, M. I., & Zepka, L. Q. (Eds.). (2020). *Pigments from microalgae handbook*. Springer.
15. Kumar, B. R., Mathimani, T., Sudhakar, M. P., Rajendran, K., Nizami, A. S., Brindhadevi, K., & Pugazhendhi, A. (2021). A state of the art review on the cultivation of algae for energy and other valuable products: Application, challenges, and opportunities. *Renewable and Sustainable Energy Reviews, 138*, 110649.
16. Maadane, A., Merghoub, N., Ainane, T., El Arroussi, H., Benhima, R., Amzazi, S., ... & Wahby, I. (2015). Antioxidant activity of some Moroccan marine microalgae: PUFA profiles, carotenoids and phenolic content. *Journal of Biotechnology, 215*, 13-19.
17. Mendes, M. C., Navalho, S., Ferreira, A., Paulino, C., Figueiredo, D., Silva, D., ... & Speranza, L. G. (2022). Algae as food in Europe: An overview of species diversity and their application. *Foods, 11*(13), 1871.
18. Mohamed Abdoul-Latif, F., Ainane, A., Achenani, L., Merito Ali, A., Mohamed, H., Ali, A., ... &Ainane, T. (2024). Production of fucoxanthin

from microalgae *Isochrysisgalbana* of Djibouti: Optimization, correlation with antioxidant potential, and bioinformatics approaches. *Marine Drugs, 22*(8), 358.

19. Mohamed Abdoul-Latif, F., Ainane, A., Houmed Aboubaker, I., Merito Ali, A., Mohamed, H., Jutur, P. P., &Ainane, T. (2024). Unlocking the green gold: Exploring the cancer treatment and the other therapeutic potential of fucoxanthin derivatives from microalgae. *Pharmaceuticals, 17*(7), 960.
20. Pereira, L. (2021). Macroalgae. *Encyclopedia, 1*(1), 177-188.

MIX
Papier aus verantwortungsvollen Quellen
Paper from responsible sources
FSC® C105338

Printed by Books on Demand GmbH, Norderstedt / Germany